stellar dust grain: diameter 4×10^{-5} inch

Blue light wavelength: 1.9×10^{-5} inch

Bacterium: diameter 4×10^{-5} inch

hole: diameter 40 miles

Large moon crater: diameter 120 miles

Largest asteroid: diameter 620 miles

diameter 4,217 miles

White dwarf: diameter 8,000 miles

Venus: diameter 7,521 miles

THE THIRD PLANET

Other Publications:
AMERICAN COUNTRY
THE THIRD REICH
THE TIME-LIFE GARDENER'S GUIDE
MYSTERIES OF THE UNKNOWN
TIME FRAME
FIX IT YOURSELF
FITNESS, HEALTH & NUTRITION
SUCCESSFUL PARENTING
HEALTHY HOME COOKING
UNDERSTANDING COMPUTERS
LIBRARY OF NATIONS
THE ENCHANTED WORLD
THE KODAK LIBRARY OF CREATIVE PHOTOGRAPHY
GREAT MEALS IN MINUTES
THE CIVIL WAR
PLANET EARTH
COLLECTOR'S LIBRARY OF THE CIVIL WAR
THE EPIC OF FLIGHT
THE GOOD COOK
WORLD WAR II
HOME REPAIR AND IMPROVEMENT
THE OLD WEST

This volume is one of a series that
examines the universe in all its aspects,
from its beginnings in the Big Bang to the
promise of space exploration.

VOYAGE THROUGH THE UNIVERSE

THE THIRD PLANET

BY THE EDITORS OF TIME-LIFE BOOKS
ALEXANDRIA, VIRGINIA

CONTENTS

OUT OF DARKNESS

More than five billion years ago, a vast, diffuse cloud of gas and dust drifting in the Milky Way was hit by a powerful shock wave, perhaps the explosive blast that marks the death of a massive star. Slammed together by this cosmic fist, the cloud's far-flung particles—mostly hydrogen and helium, with traces of heavier elements left over from other stellar deaths—concentrated their mutual gravitational attraction, causing the cloud to begin contracting. Random bits of turbulence turned into slow eddies and whirlpools that broke off on their own. As one of these cloud fragments continued to condense and contract, it began spinning faster, as skaters do when they pull in their arms. Gradually, the gaseous sphere spread itself into a thin disk 50 billion miles wide, the nebulous ancestor of the Solar System.

A thousand years passed: Heavy elements—iron, nickel, and silicon—fell toward the center of the Solar nebula, heating it even as the disk's farther regions cooled. Small particles collided, stuck together, and became bigger particles, drawing the nebula into increasingly dense strands and eventually into planetesimals, small bodies a few miles in diameter. Over a much longer time, the planetesimals, circling a growing central body, or protostar, snowballed into planetary embryos that swept up nebular debris in their vicinity. After tens of millions of years, the protostar held more than 90 percent of the original material of the Solar nebula. Now massive enough to fuse hydrogen atoms in its core, it ignited a continuous thermonuclear explosion—and the young Sun began to shine on its clutch of orbiting planets.

These smaller bodies were themselves far from finished, however: The third one out, for instance, was little more than a battered rock. The transformation of this barren pebble into Earth, the Solar System's garden planet, would take another two and a half billion years.

A Glowing Coal

Having shaped itself grain by grain out of stellar leftovers, the new Earth turned under a sky scribed with the trails of comets traveling on crazily eccentric paths. Many of these errant celestial bodies smashed into the cold, rocky surface of the young planet, cratering its face and adding fresh mass and the heat energy of impact.

Earlier on, a tenuous and short-lived hydrogen atmosphere had wreathed the blasted globe, but in the face of the strong solar wind—the current of charged particles streaming out from the Sun—Earth's gravity was unable to hold on to the volatile gases. Most of the Solar nebula's original supply of hydrogen and helium had long ago been flung toward its outer reaches, partly by centrifugal force, partly by the solar wind; in time, these would coalesce into an array of giant gaseous planets.

For tens of millions of years, Earth showed no vital sign. The barren planet simply whirled silently in its orbit, its crust cracking and shrinking as the surface cooled to the deep chill of space. But even as the outermost layer of the globe seemed to wither, forces were at work within. The contraction of the crust began to compress the minerals in the planet's core, causing temperatures there to rise. As shrinking continued, the interior grew ever hotter, until the hidden store of metals began to melt, the start of a process that would eventually quicken Earth to geophysical life.

GIFT FROM A STAR

Outwardly quiescent, Earth hid an active interior beneath its crustal shell. Mixed in with the jumbled silicates and metals of the planet were scattered remnants of an ancient supernova—heavy elements such as uranium and thorium, whose massive atoms can be forged only in the heart of an exploding star. Such atoms are unstable, and as they gradually decayed in Earth's heart, they released high-energy particles and radiation in the form of heat.

Combined with the heat generated by the relentless impact of meteoroids and the globe's continued gravitational compression, the radioactivity began to liquefy the rock beneath the crust, causing its constituents to separate. Gravity pulled iron and nickel down toward the center to form a primitive core. The white-hot slag of less dense, nonmetallic silicates floated upward, forming an overlying sphere of fluid rock now known as the mantle. The radioactive elements, whose large atoms are not easily accommodated in tightly structured metals like iron, were carried up with the silicates, continuing to generate heat as they decayed.

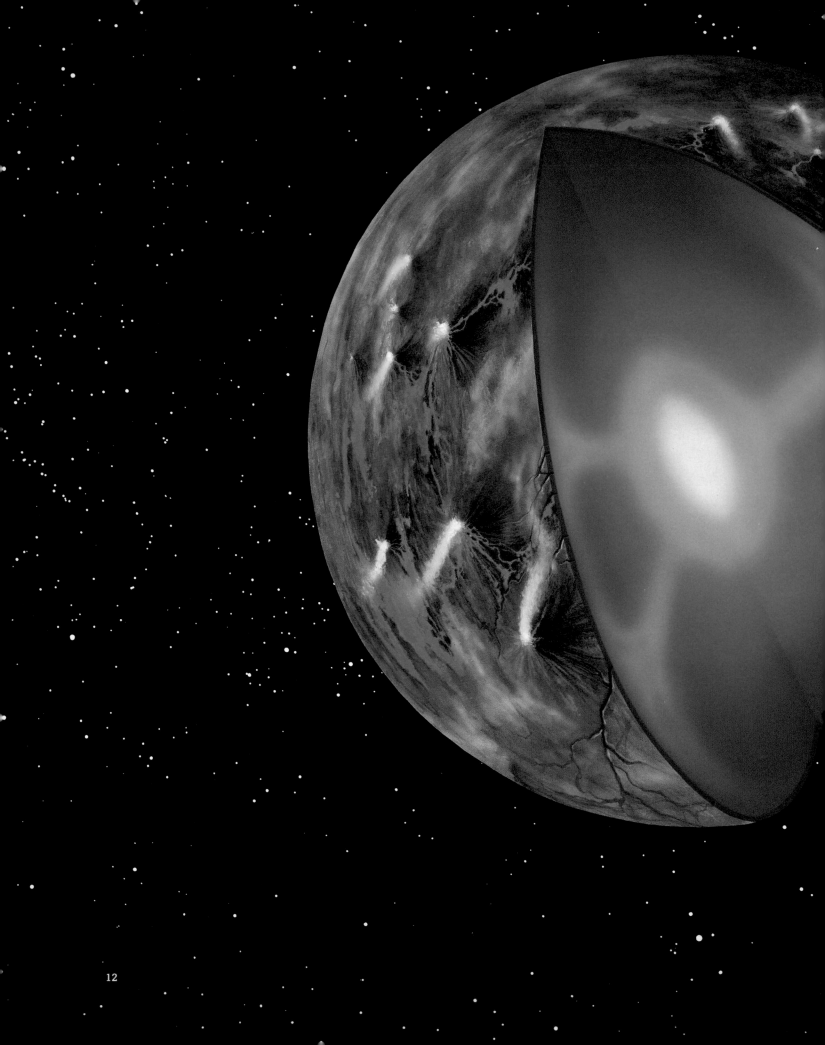

Fire and Brimstone

The release of gravitational energy as metals sank toward Earth's core and the continuing radioactive decay of heavy elements produced enough heat to melt the downward-drifting iron. This flow of heat caused convection currents to rise and sink in the fluid rock, or magma, of the mantle. Pressed by upsurging magma, the crust began to give way at its thinnest points, and for the first time lava spewed onto the surface, dotting the landscape with the rough cones of volcanoes and with ponds and lakes of liquefied rock. The primordial crust disappeared as the fiery deluge smoothed the pocks and wrinkles of a world turning one billion years old.

Just as stirring cools a steaming bowl of soup, the slow circulation of fluid rock cooled the planet, and a new crust congealed over the mantle. Deep within, the material in the core separated further as gravity compressed the innermost metals back into solids, leaving an outer core of molten iron. Convection in the hot, fluid iron turned Earth's heart into a dynamo that began to generate a current of electricity. As a result, a magnetic field spread spaceward, sheathing the planet in invisible lines of force, a barrier against the high-energy particles pouring from the Sun.

Out on the surface, another transformation was beginning as the vented lava brought with it abundant gases cooked out of the planet's very rocks: Rising from Earth's molten depths into frigid space, the gaseous mixture began to build the young planet's primitive atmosphere.

Fashioning an Atmosphere

Locked in the silicon compounds of the mantle, water made its terrestrial debut as steam, vented by the fiery volcanoes punching through the crust. More water was forged in high-temperature interactions between hydrocarbons and oxygen imprisoned in molecules of silicates and iron oxides. Emerging into the cold of space, water vapor condensed and fell back to the planet, forming puddles and blurring the edges of craters left by ancient bombardments.

Even as rain changed and sculpted the Earth's face, another by-product of hydrocarbon reactions appeared: carbon dioxide. Because CO_2 is transparent to incoming short-wave radiation but absorbs the longer-wave heat emanating from the planet itself, waxing and waning supplies of this critical compound in the atmosphere and crust would help in the shaping of Earth's evolution.

A third player in the terrestrial atmosphere was nitrogen, whose presence seems to be the result of a cosmic case of mistaken identity. During the formation of the ancestral Solar nebula, molecules of ammonia, which are composed of nitrogen and hydrogen, sometimes accumulated in the lattices of silicate compounds, taking the place of atoms of potassium, which are nearly the same size. In the fierce crucible of planetary formation, almost all the planet's nitrogen was set free to become the dominant element in Earth's atmosphere.

RAIN AND A GLOBAL OCEAN

Beneath weeping skies, rudimentary basins of water grew into a global sea. By the time Earth was nearly two billion years old, the planet had secured a unique status in the Solar System. It was now an oceanic world, its surface carved and mostly covered by a thick skin of water, overlain in turn by a gauze of atmosphere. Carbon dioxide, once a prominent atmospheric gas, had begun to be absorbed into the upper layers of the sea and, through geologic processes, into calcium and magnesium carbonate rocks.

The solid surface, too, was transformed. As the crust cooled, thickened, and cracked, it broke into a mosaic of huge plates. Moved by the hot currents below, the plates slowly rearranged themselves as the eons passed, crunching together, drifting apart, and opening vents to allow magma to rise from the depths to form new crust.

As continents rose and sailed away on these giant rafts, only to collide in another age, the early ocean gradually became an organic nursery. Somehow, certain carbon-based molecules grew in complexity, replicated themselves, and evolved to the condition that would be known as life.

Behold: Earth

For half a billion years, the simple, microscopic life forms propagating in Earth's oceans were sustained by the organic molecules abounding there. Then that supply diminished, and the tiny organisms began to feel the grip of famine. Gradually, however, they evolved a way to put the Sun to work. In a process called photosynthesis, they began to employ sunlight to extract hydrogen from water and carbon from carbon dioxide. Using hydrogen in the reaction, the organisms—still far less complicated than a bacterium—exploited the carbon to build new tissue. The oxygen in the two original compounds was used as a reactant, then discarded as waste.

As the creatures spread through the oceans and continued to evolve, their respiration pumped ever more free oxygen into the environment—a death knell for life forms accustomed to an anaerobic, or oxygenless, existence. For the algae and other green plants that had adapted, however, life was just beginning.

Still, living things kept mostly to the sea, protected by the water from the Sun's powerful ultraviolet radiation. Then, high in the atmosphere, another shield began to form. Atoms of molecular oxygen, ripped apart by the ultraviolet rays, recombined to form an unstable, three-atom molecule of oxygen called ozone, absorbing the destructive radiation in the process. Ozone concentrations increased, creating a patchy layer some fifteen miles above the planet's surface. Under its protection, organisms at last began to venture onto land, the next step on the long journey to the present, more than two billion years away.

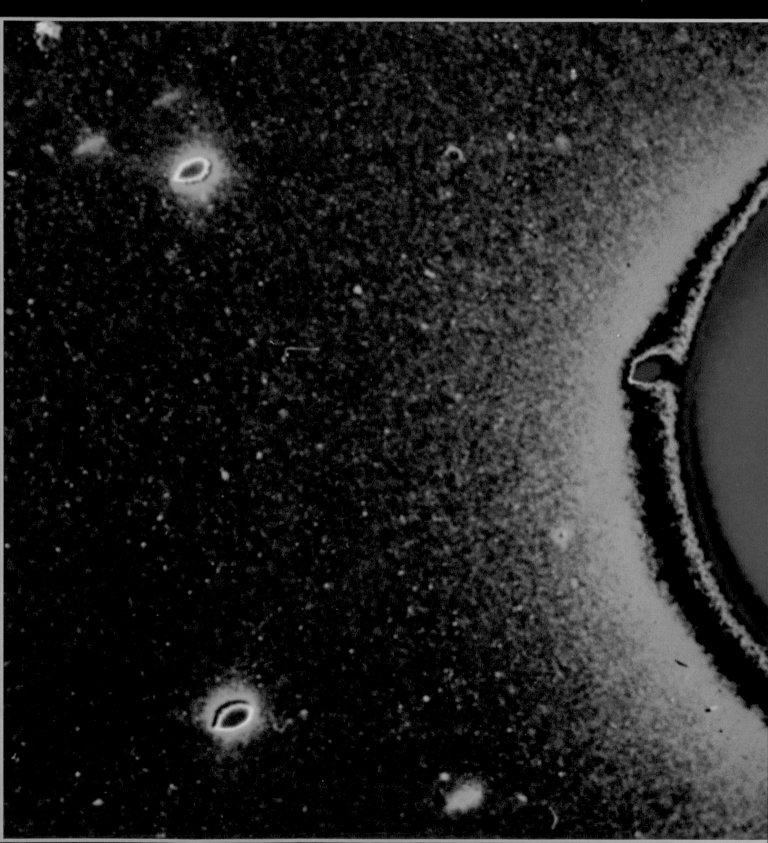

Bombarded by energetic ultraviolet photons from the Sun, atoms of oxygen and nitrogen in Earth's atmosphere fluoresce at different brightness levels, depending on their concentration and exposure to sunlight. The brightest area, designated red in this color-enhanced photograph taken during a lunar mission in 1972, shades to yellow at higher altitudes and at the day-night boundary, and then to blue on the night side and in the farthest reaches of the atmosphere

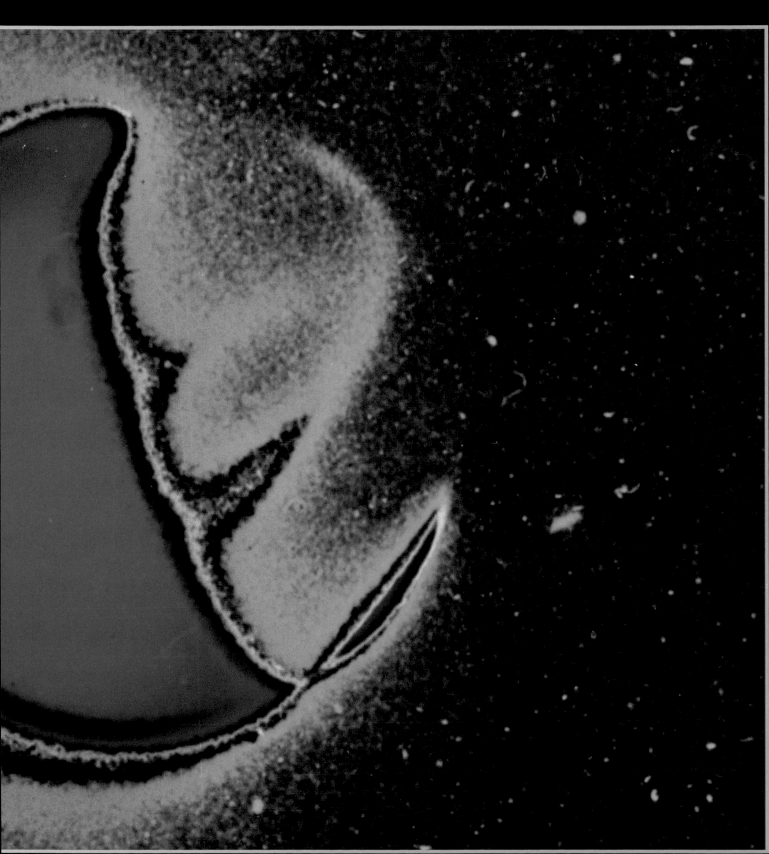

he great disk of dust and gas that spun the Sun into existence billions of years ago produced much else besides—planets, moons, meteoroids, asteroids, and comets, all still wheeling in the glow of the central star. Near the hub of this celestial dance, four hard-shelled planets trace their rounds. They are kin in many ways, fashioned of the same basic stuff. But they are also very different worlds. Mercury, the smallest, most sunward, and least protected of them, is as barren as a stone, its scorched and cratered face an open record of the meteoritic battering it has received over the eons. The other three, veiled by atmospheres, seem to promise more vitality. With one exception, however, the promise is false. On shrouded Venus, 67 million miles from the Sun, solar heat is trapped and amplified by a thick greenhouse of carbon dioxide, producing surface temperatures in excess of 800 degrees Fahrenheit. On Mars, some 75 million miles farther out, daytime temperatures rarely rise above freezing, and the atmosphere is too thin to trap heat or offer much protection against the barrage of solar radiation that sterilizes the planet's very soil. The two planets' extreme temperatures preclude the existence of liquid water, with all its extraordinary properties *(pages 24-25):* On Venus it turns to vapor, on Mars to vapor or ice.

Earth, the third planet from the Sun, is the exception. There, several layers of atmosphere form an effective sunscreen and help maintain a delicate balance between incoming radiation and heat reradiated from the surface. Clouds made of liquid water and ice form bands and whorls as they respond to Earth's rotation and to changes in heat and atmospheric pressure in the gaseous nitrogen-oxygen mixture known as air. The oceans below—covering more than two-thirds of the planetary surface—are also in complex motion. Together with the atmosphere's winds, ocean currents balance equatorial heating against the perpetual winter of the poles, storing and transporting the solar energy that governs surface seasons of rain and drought, fishing and farming. In addition, the ocean regulates the atmosphere by exchanging warm surface water and the chilled water of its deeps over long intervals of time.

The ceaseless circulation of air and water is only one sign of Earth's vigor. The seemingly solid, slightly flattened sphere is likewise animated. Not quite 8,000 miles in diameter, it contains six and a half billion trillion tons of rocky material, more than a third of it iron, more than a quarter oxygen, 17 percent magnesium, 13 percent silicon, and 7 percent everything else. Its deep interior is hot and liquid, a dynamo that generates a sweeping magnetic field around the planet. Closer to the surface, the slow movement of plastic rock contributes to processes that build islands and move continents, continually transforming the face that Earth presents to the universe.

None of these systems acts in isolation; all are intricately linked to each other and to the biosphere, the planet's slender membrane of life. In the solar family, Earth may well be the only home to living matter: Several trillion tons of it are distributed among one and a third million species of animals and plants, including bacteria that drift on stratospheric winds, microbes that live as much as several miles deep in the solid Earth, and eyeless creatures that inhabit the deep ocean trenches. Like minute cogs in a massive machine, the births and deaths of the myriad life forms—indeed, their very inhalations and exhalations—help drive the planet's constant cycle of decay and renewal.

THE HOUSE OF USSHER
Wind, water, rock, and life all beat to an underlying orbital pulse as Earth spins through its days and circles the Sun in the course of its year. For centuries, these diurnal and seasonal tempos were all that humans knew of time. A species with a life span of perhaps fourscore years could best comprehend great stretches of time in terms of the passage of generations; humans had no inkling of how very old the world they inhabited was.

One early attempt to calculate Earth's age relied on revelations in the Book of Genesis. Basing his figures on the genealogy set forth in the Old Testament, James Ussher, a Dublin-born Anglican archbishop, proposed in 1650 that Earth was created at 9 a.m., October 23, in the year 4004 BC. The first humans appeared on October 28, and Noah's deluge followed 1,655 years later, in 2349 BC.

Although Ussher's estimate was widely adopted, especially by biblical scholars, scientists grew increasingly troubled by the modest size of this canvas. The planet's surface was observed to alter only very slowly; the layering and folding and sculpting of rock required ages. Contributing to scientists' unease with a planet only sixty centuries old were discoveries of the fossil imprints of sea creatures in mountain rocks, many miles from the shoreline. The implications were disconcerting: If mountains had once been under the sea, either the land or the ocean had moved, and no one knew of a process that could account for such movement, certainly not over relatively short periods of time. Moreover, other fossil evidence seemed to indicate that some animal species had simply vanished from the face of the planet—an occurrence that suggested the fragility of all life. Reluctant to abandon Ussher's closely reckoned chronology, however, and seeking an explanation for extinctions, scientists invented an explanation that came to be called catastrophism.

According to this theory, Earth led a generally untroubled existence for most of its past. But now and then it was visited by a catastrophe of global proportions—a great flood, perhaps, or a spasmodic bout of earthquakes and

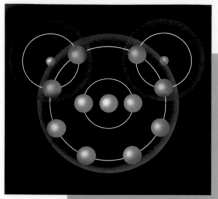

ELIXIR OF EVERYTHING

Earth has sometimes been called the Water Planet—and for good reason. Present-day Earth contains some 370 trillion trillion gallons of the compound, most of it in the oceans and polar icecaps. Although simple in its makeup, water is a chemical maverick. For example, the reason ice floats is that water, unlike almost every other substance, expands when frozen rather than contracting, decreasing in density rather than growing more dense. Were this not the case, Earth would long ago have frozen solid, from the bottom of its oceans and lakes up.

Most of water's unusual properties may be attributed to its molecular structure and to the strength of the bonds that form between its two atoms of hydrogen and solitary one of oxygen. The orbital shell surrounding the nucleus of a hydrogen atom has space for two electrons but contains only one. The shell of an oxygen atom, with room to accommodate eight electrons, holds only six. In this unfilled state, both kinds of atom are unstable, with their electrons barely held in place. But in forming water, oxygen fills its empty places with the single electrons of two hydrogen atoms, and each hydrogen atom in turn fills its shell with one of the oxygen electrons. These covalent bonds, as they are called, produce a stable molecule that bears a faint resemblance to the head of a cartoon mouse, with two hydrogen atom ears set 105 degrees apart on the oxygen head (far left).

The slightly lopsided configuration accounts for several of water's characteristics. The hydrogen side of the molecule carries a positive electrical charge and the oxygen side carries a negative one, making each molecule what is termed a dipole—the electrical equivalent of a tiny magnet. When water molecules interact with substances such as sodium chloride (table salt), whose atoms are held together by simple electrical attraction rather than by covalent bonds, the dipole effect can be very disruptive, canceling out some of the attraction, separating the oppositely charged atoms, or ions, and making room for more water. As the ions are surrounded by water molecules, the substance dissolves. Water is so nearly a universal solvent that even a sip from a glass of water contains minute quantities of dissolved glass molecules.

The dipole effect also explains the phenomenon called capillarity, the tendency of liquids to rise along the surface of a solid, seemingly in defiance of gravity. In narrow tubes such as blood vessels or the roots of

volcanic eruptions of almost unimaginable intensity. Some catastrophists went so far as to postulate exactly twenty-seven such events in Earth's past. Each disaster marked the end of one sort of natural experiment and the beginning of another. The current inhabitants of Earth were presumed to have sprung from a final and permanent creation about 6,000 years earlier. Noah saved most of these new lines of terrestrial life from yet another catastrophe—but not all: Some vanished species seen as fossils were thought to be flood victims that had no representation in the ark.

Although catastrophism seemed to square with both observation and ideology, the gentleman-geologists of the late 1700s began to note that the fit was far from perfect. The sticking point continued to be the sluggish pace of change in the natural landscape. Even human works testified that stones, once set in place, are not easily rearranged. For example, the wall strung across the northern counties of what is now England by the second-century Roman emperor Hadrian had not changed much in 1,600 years, nearly half the time since the presumed date of Noah's flood. If Hadrian's trees, for example, the tubes themselves pull liquid upward by forming hydrogen bonds with water molecules. The water molecules, in turn, pull on one another through the same mechanism.

Fortunately for terrestrial life, water's strong hydrogen bonds give rise to another unusual property. The amount of heat needed to raise the temperature of a given quantity of water a given number of degrees—known as heat capacity—is quite high, in comparison with other substances. If water at sea-level pressure behaved in the same way as its chemical cousins—such as H_2S, or hydrogen sulfide—it would freeze at minus 148 degrees and boil at minus 132 degrees Fahrenheit. At the range of temperatures that has existed on Earth for most of its life, all liquid water would have turned to steam. Instead, the hydrogen bonds make water freeze sooner than compounds of similar molecular weight but also prevent it from heating quickly. A pot of water requires so much energy to reach the boiling point that, for example, the metals in the pot handle will get hot enough to burn an unwary hand while the water itself is still cool. Moreover, the same amount of energy is needed to melt ice. In other words, ice cubes cool a drink by drawing heat from the surrounding liquid as they melt.

Because of water's tremendous capacity to absorb heat, a molecule of water vapor carries a large cargo of stored, or "latent," heat energy with it. When vapor cools enough to condense into a liquid, that heat is released to warm the surroundings, a phenomenon that drives the planet's weather and climate *(pages 85-95)*. The atmosphere's relatively rapid cycles of evaporation and condensation are braked by the oceans covering more than two-thirds of Earth's surface: The hydrogen bonds that make water slow to warm allow this enormous reservoir to regulate the transfer of heat around the globe.

That water exists in liquid form on Earth is clearly the luck of the cosmic draw. The planets to either side are examples of alternate fates: If Earth had formed in Venus's orbital position, just 10 percent closer to the Sun, the oceans would have evaporated long ago; if it had formed where Mars orbits, only 10 percent farther away, every drop of the planet's water would be locked in eternal ice. Instead, Earth has managed to retain virtually all of its original endowment of liquid water. Although water evaporates into the atmosphere at an estimated rate of 123,000 cubic miles each year, almost none of this is lost to space. Rather, water molecules spend an average of ten days on an aerial journey of several thousand miles before eventually returning to Earth to start the cycle again. Conceivably, the water molecules in the morning coffee may once have slaked the thirst of dinosaurs or fallen as rain on the armies of Alexander the Great.

Wall had existed undisturbed for so long, the terrain it crossed must have remained essentially free of major upheavals for at least that many centuries, and perhaps for many more.

NEPTUNE'S LAYER CAKE

Catastrophism was invented to preserve Ussher's estimate of Earth's age, but the planet refused to bow to the limitation. Scientists continued to look for ways to resolve the differences between belief and observation. One hypothesis was advanced in the 1780s by Abraham Gottlob Werner, a professor of mining and mineralogy in the village of Freiburg, about twenty miles southwest of Dresden in what is now the German Democratic Republic. After several years at the University of Leipzig, Werner joined the Bergakademie Freiburg in 1775. Over the next four decades, his efforts would transform the little school into a European center of excellence in mineralogy.

From his provincial academic base, Werner studied the geologic formations he found close to home. These were rather neatly stratified: a granite foundation overlain with successive deposits of limestones, clays, and sands. The beds of stone were penetrated by ore-bearing fingers of crystalline rock, which cracked the landscape into a mosaic of many-layered blocks. The province of Saxony drew a thriving coal and mineral business from this ordered setting. Werner drew from it what came to be called a Neptunist theory of the Earth.

He believed that the Earth began as a deep primordial ocean enveloping a rather small, irregularly shaped nucleus of solid material. Over time, minerals suspended in this universal fluid rained down upon the nucleus to build up as different types of rock. The granites and other crystalline rocks were the first and oldest, followed by slates and some limestones, then sandstone, coal, and basalt, and finally, sands, clays, and gravels. As each series formed, the ocean level fell, exposing the new rock and providing a site for life on land. The fact that rocks do not occur everywhere in the same sequence was deemed the result of a resurgence of the global ocean and of what Werner called local effects.

Because Neptunism invoked worldwide floods, catastrophists were comfortable with the theory, but other thinkers doubted the likelihood that all of Earth's mass of rock could have been suspended in Werner's primordial ocean. The crucial challenge to Neptunism came from James Hutton, whose lifelong field of view had been not the rolling, stratified terrain of southern Germany but the ice-harrowed valleys and stony hills of Scotland.

A GRANITE LABORATORY

In 1750, following a broad education in Britain and France and a brief medical practice in London, the twenty-four-year-old Hutton returned to his native Edinburgh. During his time on the Continent, he had acquired an abiding fascination with geology and, as a result, had decided to abandon medicine. In its place, he took up farming. The daily work on his Berwickshire property some forty miles south of the city offered him ample opportunity to study the details of the countryside. It also brought him prosperity, and by 1768 he was able to rent out his land and move to Edinburgh.

For the next seventeen years, amid a landscape clawed by ancient glaciers and pierced by jutting crags and the eroded calderas of five extinct volcanoes, Hutton polished a radical hypothesis. Finally, in 1785, he read "A Theory of the Earth" before the Royal Society of Edinburgh.

His ideas were as simple and direct as his paper's title. The true story of creation, Hutton argued, lay not in any human record but in what he called natural philosophy—the tale written in the soil and rocks of the planet. In his view, that record reflected endless self-renewal rather than the convulsive rhythms of catastrophism. He reasoned, for example, that sedimentary rocks like sandstone were formed from material washed off the continents by water and wind and deposited layer by layer at the bottom of the sea, and that they were somehow then uplifted from their watery birthplace. Hutton also theorized that granite, rather than being a primeval material, was in fact a more recently molten rock, thrust upward through the Earth's crust from an interior of unimaginably high temperatures and pressures.

More important than the process, however, was the vast stretch of time Hutton sensed behind the geologic machinery. "We cannot estimate the duration of what we see at present, nor calculate the period at which it had begun," he wrote. "With respect to human observation, this world has neither a beginning nor an end."

JUPITER'S TEMPLE

In effect, Hutton presented scientists with a canvas adequate to the task of describing the planet and its life—but few of them had the confidence to use it. The biblical time scale of 6,000 years remained essentially unchallenged. Even the most innovative of Hutton's contemporaries were timid about extending the boundaries of geologic time, proposing spans that ranged from 75,000 to a few hundred thousand years since Earth's formation. It fell to another Scot to make the case for an Earth that was hundreds or thousands of times older.

Charles (later, Sir Charles) Lyell was born in 1797, the year of Hutton's death, and demonstrated an early interest in the field of geology while still at Oxford's Exeter College. At the age of twenty-two, he set out on a long continental journey for the purpose of resting his study-weakened eyes. As it happened, the trip through the Rhine Valley and the volcanic lands of southwestern Italy honed his geologist's vision. Returning to the areas seven years

With Australia and the Pacific peeking through the cloud cover, the blue orb of Earth noses above the lunar surface in an image taken during the mission of *Apollo 11.* Shot from the command module *Columbia* during one lunar orbit, this and the photographs on the following pages record earthrise less than twenty hours before the first explorers from the Third Planet set foot on the Moon.

later, he visited extinct volcanoes in the French Auvergne and went again to Italy to meditate on the region around the grand volcanoes Aetna and Vesuvius. It was in the shadow of Vesuvius that Lyell found what would become his paradigm of change.

Just seaward of the city of Naples stands the ancient temple of Jupiter Serapis, a jumbled ruin marked by a row of slender marble columns. In the second century AD, according to contemporary accounts, the structure had stood well above the high-water line. At the time of Lyell's visit, the foundation lay a foot beneath that mark—but the columns showed scars left by the borings of marine mollusks to a height of more than twenty feet. The implication was clear: Sometime since the second century, the temple had sunk beneath the Bay of Naples to at least that depth and then risen again to its current position. Lyell inferred from this not only that the land had moved vertically but that the fall and rise had been gentle enough to leave the fragile columns still upright sixteen centuries later.

To the young scientist, the temple buttressed Hutton's thesis and his own growing convictions: There was no need to invoke catastrophes or processes more intense than those currently observed to explain the world as it is. Rather, the Earth is eternally eroded and renewed by the same forces over great intervals of time. In 1830, Lyell presented his theories in *Principles of Geology,* a work of such scope and power that it transformed what had been largely the province of natural historians into a modern science.

As the science matured, geologists accumulated further clues to the massively slow pace of change on the planet. Such change could be seen, for example, in structural dislocations that had pushed huge slabs of rock over neighboring formations for tens of miles. They also learned that Earth's climate had fluctuated dramatically from epoch to epoch: By studying large areas covered by sand and gravel, they deduced that such fluctuations caused ice sheets to spread periodically across great expanses of the northern continental mass. But even as the notion of an indescribably ancient Earth began to take hold, another scientist began to whittle away at the planet's age with formidable instruments: the unyielding laws of physics.

THE SPOILER

William Thomson—later Baron Kelvin of Largs and best known today for the Kelvin absolute-temperature scale—was a Belfast-born scholar who came to personify turn-of-the-century physics in Great Britain. In 1860, the thirty-five-year-old Kelvin was teaching at the University of Glasgow. The scientific issue that most intrigued him at the time was whether the Sun's energy derived from mere chemical oxidation, the kind of fire common on Earth. If so, the Sun had enough mass to burn for about 10,000 years, no more. Kelvin subscribed to another theory, however. He believed that the Sun got its energy from the gravitational in-falling of meteoric bodies (in the act of falling, the material's potential energy was converted to heat) and from the heat generated as gravity compressed the star. By carrying out a series of calculations involving the mass of the Sun and the intensity of its radiation, he arrived at a figure for its age: about 20 million years.

Given this outer time limit, Kelvin then proceeded to the problem of the age of the Earth. Even as a student at Cambridge in the 1840s, he had distrusted the idea of an Earth with neither beginning nor end, finding the key to his opposition in the laws of thermodynamics, which describe the conservation and dissipation of heat. Now he argued that Earth—whether it had been torn out of the Sun or formed along with it—came into existence as a ball of molten material. The heat generated by the material's inward gravitational collapse would have dissipated rapidly at first, and then more slowly, as the temperature of the nascent planet approached that of its surroundings. Earth's surface was obviously considerably cooler now, but one had only to visit a deep mine to know that the globe's interior was still hot. Kelvin measured the way temperature increases with depth and the thermal conductivity of the silicate rocks known to dominate the outer

crust. Factoring in a calculation of the rate at which the planet would radiate its heat into space, he determined that Earth was somewhere between 25 million and 400 million years old.

Over the next three decades, various geologists proposed other schemes for gauging Earth's age; all were based on known and measurable geologic processes, but none could muster the weight of Kelvin's numbers and laws. In 1897, after additional refinements, the seventy-three-year-old scientist issued a final calculation: Earth could have been habitable for only 20 to 40 million years. There the matter might have rested, except for a chain of apparently unrelated discoveries that had begun across the English Channel the year before.

COSMIC ROCKS

At forty-three, Antoine Henri Becquerel, having given up active research, was a successful teacher of physics at the Conservatoire National des Arts et Métiers, the Museum of Natural History, and L'École Polytechnique. But news of the discovery of x-rays by the German physicist Wilhelm Roentgen tempted him back to the laboratory in early 1896. Over the next few years, Becquerel carried out a modest study of minerals that emitted penetrating rays capable of exposing a photographic plate even in complete darkness. Potassium compounds produced some mild emissions, but a disk of pure uranium, he found, produced three times as much. He labeled the phenomenon "phosphorescence of a metal." His friends and colleagues, Marie and Pierre Curie, called it radioactivity and shared the 1903 Nobel prize with Becquerel for the discovery.

That same year, the Curies reported another finding, which they could not explain: Radium was warm to the touch. Across the Atlantic, two young researchers at Montreal's McGill University had also noted this mysterious heat radiation and thought they understood its cause. Transplanted New Zealander Ernest Rutherford, at twenty-nine fast becoming the preeminent experimental physicist of his day, and twenty-three-year-old Frederick Soddy, a brilliant Oxford-trained chemist, believed that the heat was a by-product of a steady breakdown of the atoms of elements like radium and uranium, and that the same process generated Becquerel's penetrating rays. They were right. The so-called rays were streams of particles, labeled alpha particles by the two researchers. These turned out to be the positively charged nuclei of helium atoms consisting of four elementary units: two protons, carrying a positive electrical charge, and two uncharged neutrons. Because an element is defined by the number of protons in its atom—its mass is the sum of the atom's neutrons and protons—losing an alpha particle transmutes the element into something else. In the argot of the physics laboratory, the original element is the parent and the elements it becomes are daughters.

The idea that heat could issue from radioactive decay, as this process is called, led British physicist Robert J. Strutt to propose that the breakdown

of radioactive elements might account for the flow of heat from Earth's interior, a notion quickly taken up by a number of researchers, including Rutherford, who speculated that a similar phenomenon might also fuel the Sun. Although no one yet understood the thermonuclear fusion process at the fiery hearts of stars, radioactive heating could keep a star burning evenly for billions of years, a steadier and longer-lived beacon than that proposed by Lord Kelvin.

In May 1904, Rutherford was invited to read a paper supporting this view before London's Royal Institution. To the New Zealander's dismay, Kelvin himself was in the audience. Rutherford did not especially want a public quarrel, and luckily, Kelvin fell asleep right away. However, as Rutherford recalled later, "As I came to the important point I saw the old bird sit up, open an eye, and cock a baleful glance at me. Then a sudden inspiration came, and I said Lord Kelvin had limited the age of the earth, provided no new source was discovered. That prophetic utterance referred to what we are now considering tonight, radium! Behold! the old boy beamed upon me."

Although Rutherford's quick thinking avoided a potentially contentious encounter, Kelvin never allowed himself to be persuaded of the argument. He died in 1907, arguing to the last that gravitational collapse was the only force capable of generating Earth's high internal temperatures.

A BROADER TAPESTRY

A decade later Kelvin's theories were still being defended by a handful of scientists, but the prevailing opinion was that the discovery of radioactivity had effectively added millions of years to Earth's life span. Decaying atoms of radioelements did more than stretch Earth's age—they also measured it. Each radioactive substance, Rutherford and Soddy learned, disintegrates at a unique rate. Uranium, for example, emits alpha particles at the rate of 12,200 particles per gram of uranium per second. This rate remains unaltered by heating, cooling, chemical combination, or anything else outside the nucleus of the atom. By measuring decay rates over relatively short periods of time, they could calculate what they called an element's half- life: how long it would take for a sample amount of radioactive material to decrease by half. A half-life could range from a few thousandths of a second to many trillions of years, depending on the material. Most naturally occurring uranium, designated uranium-238, has a half-life of about 4.5 billion years, which means that half of a sample decays over that period of time. Half of the remaining uranium-238 will decay over the next 4.5 billion years, half of the remainder over the next, and so on until all of the uranium atoms have been transformed into daughter elements.

In 1907, Bernard B. Boltwood, an American chemist at Yale University and a close friend and colleague of Rutherford's, began to examine the new time-piece. A few years earlier, Boltwood had shown that uranium (which proceeds through the radioactive daughter elements thorium, proactinium, radium, radon, polonium, and bismuth) ultimately ends its decay as stable lead. He reasoned that this would gradually cause the amount of uranium in a sample of rock to decline relative to the amount of lead. Measuring the ratio of lead to uranium thus would indicate the approximate age of the rock. His first efforts, which analyzed similar material from Sweden, the United States, and what is now Sri Lanka, were consistent: The rocks seemed to be from one to one and a half billion years old. With only a rudimentary feel for radioactive decay rates, Boltwood had expanded terrestrial time almost a hundredfold.

Although the theory of radiometric dating was now in place, Boltwood's pioneering efforts were only the beginning of the technological development that would be required for the procedure to become a standard laboratory tool. Eventually, however, scientists came to understand the details of radioactive decay chains and the varying rates of decay. Today, three main transmutations are used to date ancient rock samples: uranium to lead,

potassium to argon, and rubidium to strontium. All three parent elements have half-lives reckoned in billions of years.

In 1955, Clair Patterson, of the California Institute of Technology, measured the distribution of lead in oceanic sediments and arrived at the first accurate figure for Earth's age: at least 4.5 billion years. His and subsequent analyses have confirmed that the planet—though not timeless, as James Hutton and Charles Lyell had it—is old almost beyond human understanding, reaching back in time some 4.6 billion years. This figure for Earth matches corresponding results for meteorites, which are remnants of the same primordial disk of stellar debris that gave birth to the planets. Further confirmation came in the 1970s when Apollo astronauts brought back rock samples from the Moon.

Old as it is, Earth is still only middle-aged, destined to prosper for millions of centuries before its hour of doom comes. More than five billion years from now, the ten-billion-year-old Sun will balloon into a star hundreds of times its present radius, a so-called red giant. Mercury will be engulfed, then Venus. On Earth, the blazing radiation will push temperatures past the 2,000-degree mark, irrevocably transforming the garden of life into a desiccated graveyard of organic ash. Finally, the Sun's outer boundary will extend nearly to the orbit of Mars. Entangled in the tenuous body of the red giant, Earth will vaporize as it spirals toward the seething interior. Some 250 million years later, the bloated Sun will explode, leaving in its place a tiny white dwarf star and, somewhere in the darkness, a cluster of rocky cinders that were once four planets. Earth thus will end as it began, in the debris of its star.

the azure plains of India in this view from the space shuttle. The mountains mark the crumpled boundary where the Indian subcontinent smashed into Asia.

"E arthquakes for breakfast, dinner, supper and to sleep on," noted John Milne soon after his arrival in Japan in 1876. Clearly, the Earth beneath these exotically foreign islands was very different from that underlying his native Britain. But, as a professor at the new Imperial College of Engineering in Tokyo, the twenty-six-year-old geologist had no particular professional interest in the small tremors that seemed part of everyday life in Japan. It took a major earthquake that leveled buildings in nearby Yokohama in 1880 to rivet Milne's attention and draw him into the nascent science of seismology. Once in, he revolutionized it.

At first he had nothing more ambitious in mind than recording in a simple way the mysterious vibrations known as seismic waves (from the Greek *seismos,* "to shake"). Near its source, an earthquake delivers a few large and sometimes devastating shocks and a confusing medley of smaller ones that often combine to cause even more damage. As the waves travel outward in all directions, their mode of vibration can be more easily identified: Some move the earth back and forth, others from side to side and up and down. Milne and engineer colleagues developed an instrument that sensed all three sets of motions and mechanically amplified them. A scribing pen geared to a clock traced a continuous line on a paper-covered cylinder—a line that earthquakes transformed into a series of distinctive squiggles.

Over the next decade and a half, Milne amassed a unique record of earthquakes in Japan—and also of distant tremors whose vibrations sped through the planet to his seismographs. But in 1895, a disastrous fire destroyed his home, his observatory, and all of his meticulously compiled data. Dispirited, the geologist returned to Britain, but he had no intention of giving up his passionate study. He settled at Shide Hill House, on the Isle of Wight off England's southern coast. A subsurface layer of chalk made the island a particularly good site for the reception of seismic waves, and within three weeks of his arrival, Milne had two new seismographs operating. Four years later, aided by the British Association for the Advancement of Science, he had built up a network of twenty-seven instruments in countries throughout the British Empire. When he died in 1913, forty stations were operating worldwide, and Shide Hill was the seismological capital of the planet.

As recordings accumulated at these listening posts, seismologists became aware that the waves were more than the muted echoes of earthquakes: They were also probes of the planet's interior, picking out features of a world that no one would ever see directly. A seismographic record begins with a wavy line of primary waves, or P-waves, which travel much as sound waves do through air, compressing and dilating the molecules of matter in their path. Minutes or seconds later, the instrument starts recording the more ragged signature of secondary, or S, waves. Because S-waves tend to oscillate at right angles to their direction of motion, either up and down or side to side, like a length of rope being snapped, they are always slower to arrive than P-waves. P- and S-waves that hit the surface are sometimes transformed into surface vibrations that inscribe characteristic long waves on the recording.

Once seismologists learned to gauge the speed at which waves traveled, they discovered that the interval between the arrivals of P-waves and S-waves could be used to calculate the distance from a given seismograph to the earthquake's epicenter—the surface point above the underground focus, or source, of the tremor—although it could not tell them where that epicenter was. However, if distance measurements from three different recording stations were plotted as the radii of three circles with the stations at their centers, the epicenter would be found where the circles intersected.

INNER EARTH

As seismic waves journeyed through the planet, differing materials modified the waves' behavior, giving scientists hazy clues to the configuration of the hidden world. Reporting on an earthquake in Guatemala that occurred in 1902, the Irish geologist and seismologist Richard Dixon Oldham concluded that a central core within the planet was casting a kind of seismic shadow onto the side of Earth opposite the shock. Oldham did not think that the core was necessarily opaque to earthquake waves but rather that waves entering the core at an angle were bent the way light is refracted when it passes through water, a medium where light's speed is less than it is in air. Rather than continuing in a straight line through the core, most of the waves would swerve, creating a virtually wave-free sector on the planet's far side.

A few years later, after a minor earthquake shook the Yugoslavian city of Zagreb, meteorologist-turned-seismologist Andrija Mohorovičić made an intensive study of the quake's seismographic record, collecting data from twenty-nine stations as much as 1,500 miles away from the epicenter. In the readings from stations within 125 miles of the tremor, he detected what seemed to be two sets of P- and S-waves traveling at different speeds. Upon further study, he realized that the quake had produced only one set, but that at a certain depth, which he calculated to be about thirty miles down, the waves split into two groups, one of which speeded up.

Seismologists had already determined that the velocities of P- and S-waves increase with the density of the material they pass through. Mohorovičić figured that the split occurred at a boundary between Earth's outer layer, or crust, and some denser rock below. One set of waves kept going through the crust at their original speed; the other set was bent, or refracted, so that it

Journey to the Center of the Earth

Though Earth's deep interior is physically inaccessible, geophysicists can explore it indirectly with instruments called seismographs. These devices record tremors that travel out from the center of an earthquake in waves, much as ripples emanate from a pebble dropped into a pond. The waves move through the Earth by setting particles of matter vibrating, traveling slowly in regions where planetary material is somewhat elastic and rapidly through regions, usually at greater depth, where rocky matter is of a stiffer consistency.

Earthquakes generate four types of seismic waves: two kinds of surface waves, which move along the ground and cause most of the damage associated with the event, and two kinds of so-called body waves, which pass through the body of the planet. Because of their penetrating power, scientists can use body waves to chart Earth's depths. Each kind of body wave has its own vibration pattern *(below)*.

Primary waves, known as P-waves, move faster and tend to arrive at recording stations earlier than secondary, or S, waves. P-waves can negotiate both solids and liquids, but S-waves travel only through solid rock. By timing the arrival of waves from a given earthquake at recording stations around the world, seismologists are able to figure out how fast the waves were traveling at different depths, what their probable paths were, and the type of material they passed through. With the help of supercomputers, scientists can manipulate decades' worth of earthquake data, plotting information for many thousands of waves to build up a three-dimensional map of the planet's interior *(right)*.

Differences in composition between one region and another cause P- and S-waves to bend, or refract, and sometimes to vanish altogether. The patterns of reception and nonreception of these waves around the globe lead geophysicists to believe that Earth possesses a solid iron inner core *(yellow)*, which may in turn be surrounded by a fluid outer core *(tan)* of melted iron and sulfur. At the boundaries between the inner and outer cores, and between the core and the mantle *(orange)*, scientists have detected a bumpy terrain that seems to be a kind of reverse image of the planet's surface.

P-waves alternately compress and dilate particles of matter, which lets the waves travel in both liquids and solids. Such waves can thus pass straight through the core to the other side. However, differences in elasticity between regions cause the waves to bend gradually, creating shadow zones *(gray)* where surface stations pick up P-waves only faintly, if at all.

S-waves move by vibrating rock from side to side, or up and down, or both—all motions that require the resistance of a solid. Thus, since S-waves are not detected on the far side of the globe opposite an earthquake *(gray)*, scientists have deduced the existence of a liquid outer core. However, P-waves passing through the inner core may

On December 7, 1988, at 7:41 a.m. UTC (universal time coordinated), a severe earthquake rocked Leninakan, Armenia. Within minutes, P-waves *(black)* had raced through the Earth, arriving at a seismic station on the other side of the world in Washington, D.C., at 7:53 a.m. Slower S-waves *(white)* arrived at 8:04.

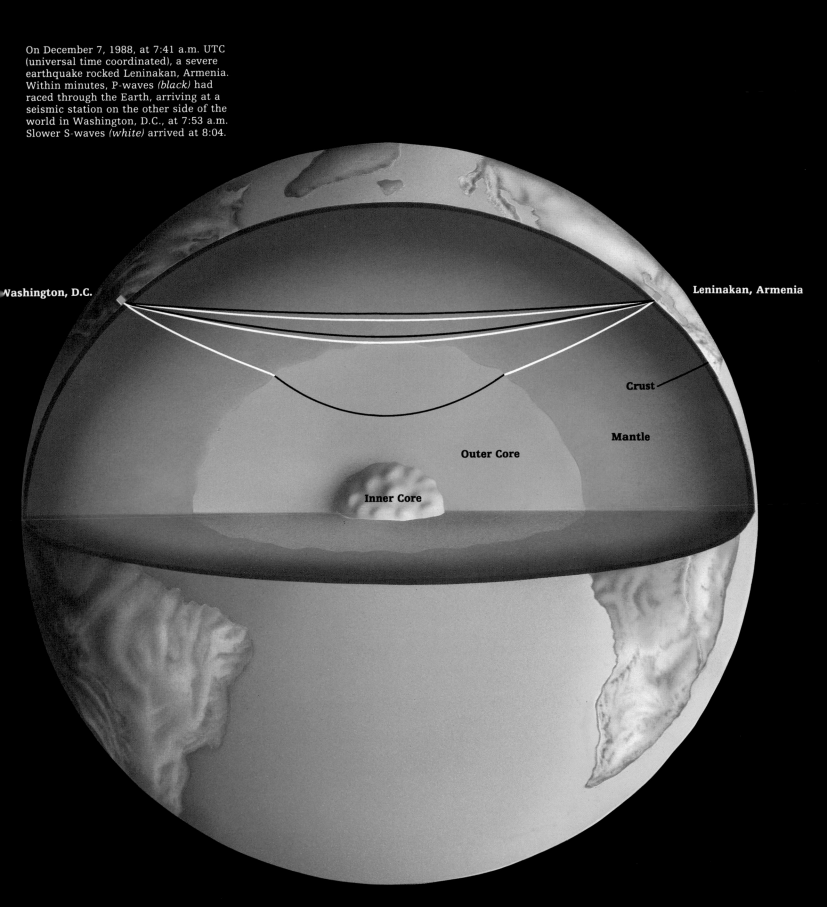

Washington, D.C.

Leninakan, Armenia

Crust

Mantle

Outer Core

Inner Core

moved along the top of the denser rock and gained speed. Thus, even though they were deeper in the Earth and had farther to travel to a given seismograph, the faster-moving waves eventually passed the waves moving in the crust and arrived first. As a tribute to his brilliant detective work, the boundary between the crust and this layer of denser rock was named the Mohorovičić discontinuity—the Moho, for short.

A MATHEMATICAL TUNNEL

Around the time Mohorovičić resolved the puzzle of the double set of seismic waves, a twenty-two-year-old graduate student at Germany's University of Göttingen was mathematically tunneling to the very center of the Earth. In what later would be hailed as a truly elegant piece of research, Beno Gutenberg wrestled with the anomalous shadow zone reported by Oldham and its hint of a distinct core deep within the planet. His approach was to create several mathematical models of Earth's interior: Positing a dense core at varying depths and calculating the behavior—the trajectories and travel times—of P-waves moving through these hypothetical cores, Gutenberg then tried to match the modeled behavior to the seismological record. By 1912, he had obtained a best fit: The mathematics seemed to indicate a boundary at a depth of 1,800 miles, between the dense rock of the region whose upper boundary Mohorovičić had discovered and an even denser core about the size of the planet Mars. In effect, Gutenberg had defined the lower limit of the layer now known as the mantle.

Gutenberg believed, as did many others, that the core must be solid. Richard Oldham thought not. He noted that P-waves, which compress and dilate matter in their path, pass through this region without difficulty but that S-waves do not penetrate beyond the core-mantle boundary. Because S-waves shake rock up and down and from side to side, they require a medium that resists such twisting. They therefore cannot traverse a liquid, which merely yields and absorbs their energy. Oldham surmised that the core was probably molten metal alloy composed mostly of iron. In this he followed a longstanding belief that only an electrically conducting metallic core that behaved a bit like a giant bar magnet could explain the presence of Earth's magnetic field. But consensus on a fluid core would not come for a dozen years.

In 1926, the British physicist Harold (later, Sir Harold) Jeffreys made a proposal based on the study of Earth tides, the barely detectable, hours-long undulations of the ground caused by the Moon's gravity. Jeffreys suggested that such tides were possible only if the planet had a fluid core, which would act as a shock absorber for the crustal motions, compensating for the mantle's rigidity. Perhaps because Jeffreys was considered the intellectual heir of the late Lord Kelvin, most seismologists, including Gutenberg, concurred. Gutenberg even detected another nonrigid region—the asthenosphere, from the Greek *asthen,* for "frail"—at a depth of about 100 miles.

A decade later, however, the work of a late-blooming Danish seismologist would show that those who had believed in a solid core were not all wrong.

Inge Lehmann worked for a Danish insurance company until 1920, when she received her master's degree in mathematics and began a scientific career at the age of thirty-two. Eight years later, she was appointed chief of the seismological department of the Danish Geodetic Institute in Copenhagen.

In the course of her work at the institute, Lehmann became intrigued by an anomaly: Although shadow zones were generally free of seismic recordings, a few weak P-waves seemed to make it through the core and to register within the shadow zone on the far side. In a 1936 paper entitled simply "P'"—P prime—she proposed that although most P-waves were refracted at the core-mantle boundary discovered by Gutenberg, some continued on and were slightly refracted by yet another boundary between material of different seismic velocities. Earth's core, she suggested, had a core of its own that was about 1,600 miles in diameter—and was solid.

With Lehmann's work, the structure of Earth's interior seemed to be clear: The planet was apparently made up of a series of rather neat, concentric spheres. Beneath a thin crust lay the shallow, slightly mushy layer of asthenosphere, part of a mantle believed to be solid rock. Below the mantle was a large outer core of molten iron alloy, and below this, an inner core of iron under such pressure that the metal had been crushed back into its solid state.

Besides explaining the behavior of seismic waves, this structure also suggested how Earth might generate its magnetic field. While molten iron is too hot to be magnetized, the action of the Earth's rotation on the fluid outer core might turn the core into an electrical generator and produce the observed geomagnetic field. Less clear was how such an arrangement affected the nature and structure of the surface. In the early part of the twentieth century, the accepted scientific wisdom regarding the planet's outermost layer was that it was rather like the skin on a drying apple. Essentially, the globe's major features—its continents and oceans—looked now as they had when they were formed; mountains, valleys, and earthquake upheavals were simply effects brought on by the shrinking of the crust as the planet's interior cooled. However, work in a variety of disciplines turned up observations and intriguing coincidences that could not be explained by such a scenario. Why, for example, should mountains be concentrated in narrow bands rather than being scattered more or less randomly and uniformly around the world? And how to account for the jigsaw congruity of the Atlantic coasts of Africa and South America, as well as remarkable parallels in the geologic and biological histories of the two coastal areas?

SHIFTING FOCUS

For Alfred Wegener, a German meteorologist who spent much of his career studying the geophysics of Greenland, these questions became the stuff of obsession. As early as 1915, in a treatise titled "The Origins of Continents and Oceans," he had proposed a radical answer.

Continental displacement, as he called his theory, began when a single supercontinent he named Pangaea broke up nearly 200 million years ago; the pieces had been drifting ever since *(page 52)*. As Wegener put it, "It is just as if we were to refit the torn pieces of a newspaper by matching their edges and then check whether the lines of print run smoothly across." The western mountains of North and South America were massive wrinkles acquired as the continents sailed through the crust at the bottom of the oceans. India's collision with Asia created the huge folds of the Himalayas. And relatively recently—only a few million years ago—Australia-New Guinea broke away from Antarctica to crunch into the Indonesian archipelago.

THE MISSING MOTIVE FORCE

Technically, Wegener's greatest problem was to explain how the gargantuan machine of continental drift worked. Because the continents' mountains thrust vertically through the underlying rock, he assumed the continents would also be able to move through the rock horizontally, reasoning that this foundation must be like pitch, which breaks when struck but is plastic enough to flow under its own weight if given sufficient time. All one needed was enough force to drive the continents through a thin layer of such material. Perhaps, Wegener speculated desperately, this came from a combination of Earth's centrifugal force and the gravitational pull of the Sun and Moon.

The argument was not a very strong one, as Wegener himself admitted. His critics, such as Harold Jeffreys, were scathing. "The physical causes that Wegener offers are ridiculously inadequate," Jeffreys told colleagues at a meeting of the Royal Geographic Society. By Jeffreys's calculation, the forces proposed by Wegener were a millionth those needed to impel continents like icebreakers through the seafloor. Continental drift had run aground on the bedrock of geophysics.

Wegener's inability to explain what drove the continents brought the German nearly to despair. As a palliative for depression, he turned back to Greenland. In April 1930, he headed an expedition to study the frozen island, which required building temporary coastal and mid-ice stations. These facilities were barely completed before the Arctic winter struck in September. When weather isolated a two-person meteorological camp 250 miles inland, Wegener embarked on an emergency mission to resupply his stranded colleagues, taking forty days' hard sledding to get there. On November 1, his fiftieth birthday, the scientist and his Greenlander guide headed back in the gathering winter night. Neither was seen alive again.

In the spring, searchers found Wegener in the shallow grave where his guide had left him, apparently dead from a heart attack. The guide's body was never found. Wegener's family, spurning a proffered hero's funeral at home, left his remains in his Greenland grave.

For nearly two decades, Wegener's benighted continental drift theory languished as most scientists pursued other avenues of investigation into the Earth's workings. In the 1930s, a few researchers began to venture into the new field of marine geophysics. Studying the oceans was many times more difficult than exploring the continents, however. Oceanographers likened

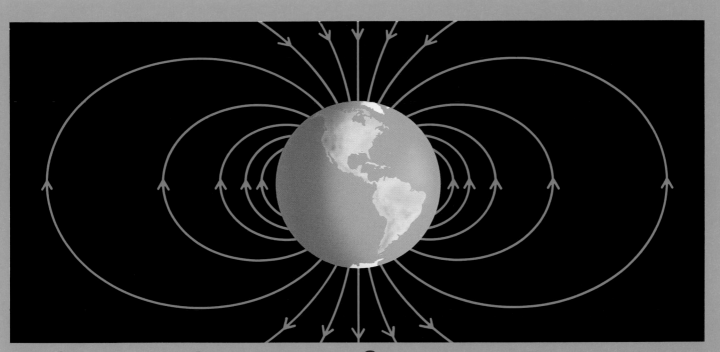

A COMPASS LOCKED IN STONE

To the amateur's eye, a rock is a rock the world over: Basalt from Japan looks like basalt from Peru. But to a geologist, magnetic minerals within such a stone identify its latitude of origin as surely as if the rock carried a birth certificate.

This fossil fingerprinting is possible because many rocks contain substances, such as iron oxides, that become magnetized as the stone cools from its molten state. In the process, the minerals align themselves like tiny compass needles along the lines of force of Earth's magnetic field. Because these field lines sweep out from the planet in vast arcs *(above)*, the minerals will point not only toward the magnetic pole but also up or down at an angle paralleling that of the field line at that latitude. For instance, minerals at the equator will be aligned horizontally; those at the north magnetic pole will point straight down.

Once the rocks harden, their magnetic pattern is fixed. In the 1950s and 1960s, geologists found that they could use these natural compasses to show how landmasses had migrated from their original latitudes. Ancient rocks in the Tibetan plateau, for instance, were apparently formed 1,250 miles south of their present location, providing convincing evidence of continental drift.

The young field of study, called paleomagnetism, has also yielded a surprising fact about the planet's magnetic field. Some rocks have been found to point not to the North Pole but to the south. Using radioactive dating techniques, researchers have surveyed samples from around the world and discovered that Earth's magnetic pole has flipped between north and south at irregular intervals at least nine times in the last 3.6 million years. The scientists do not know why Earth's polarity reverses this way—or when it will happen again.

probing the deep seafloor from a ship to exploring a new planet at night from a high-flying balloon, using buckets on long ropes to sample the surface.

One device that helped make geophysicists' lives easier was the echo sounder, developed after the sinking of the *Titanic* in 1912 to detect submerged icebergs and later adapted to hunt German U-boats during World War I. By emitting a ping and measuring how long it took for an echo to return, the instrument plumbed the water depth or showed distance to an underwater target. With sonar (for "sound navigation ranging"), scientists could record profiles of the deep seafloor from moving ships. Another invaluable technique, called seismic prospecting, had been around since the early 1920s, when petroleum geologists searching for oil began using small explosives buried just below ground level to generate artificial earthquake waves. These bounced off oil- and gas-bearing rock formations and sketched a picture of underground structure.

Maurice Ewing, a student at Rice Institute in Houston, Texas, was introduced to seismic prospecting in the summer of 1930 when he signed on with a crew looking for oil along the coastal marshes of Louisiana. After obtaining his Ph.D. the following year, Ewing taught physics at Pittsburgh and Lehigh universities, then shifted to wartime research at Woods Hole Oceanographic Institution in 1940. There, with naval funding, he took seismic prospecting to sea, probing the structure of the upper layer of sediment and underlying crust with shock waves from handmade bombs. By the end of the war, he had been named the first director of Columbia University's new geology center, named for Wall Street banker Thomas Lamont, whose family donated a New York country estate to house the new institution. (In the 1960s, following a generous endowment by the Henry Doherty Foundation, the center was renamed the Lamont-Doherty Geological Observatory.) Within a few years, Ewing persuaded Columbia to buy the yacht *Vema*, a three-masted schooner originally built in Denmark for American cereal heiress Marjorie Merriweather Post. Refitting the 202-foot vessel for geophysical research, Ewing began an unprecedented survey of the global sea.

The world revealed by Ewing's *Vema* cruises had little to do with the conventional view of the oceans. Prevailing theory suggested that the ocean basins were rather smooth, made of rock formed when Earth was born and since buried perhaps twelve miles deep in the detritus of several billion years. Ewing's profiles showed that the submerged rims of the continents—the continental shelves—were heavily laden with sediments, to a depth of several miles. But the deep ocean floors revealed a sediment

WANDERING CONTINENTS

For more than two billion years, continent-size landmasses have roamed slowly over the globe, merging and splitting apart to produce ancient agglomerations such as Gondwana *(right)*. The continental arrangements shown here and on pages 45 to 49 span a period from half a billion years ago to 150 million years in the future. The modern continents are color-coded for identification; for example, South America is brown.

514 million years ago. North America *(tan)* and parts of Asia *(light green)* lie on the equator. Australia *(pink)*, Antarctica *(dark green)*, Africa *(blue)*, and South America *(brown)* form Gondwana.

layer only a few thousand feet thick. At the presumed rate of sedimentation, the deposits could not have accumulated for more than 180 million years. In fact, subsequent dredging and coring of marine sediments failed to find any older than 135 million years.

More profound surprises were in store. Toward the middle of the Atlantic, the geologists found the sediments thinning away as the seafloor rose into a broad ridge. Oceanographers had been aware of the ridge since the early soundings made across the Atlantic by U.S. Navy hydrographers surveying possible routes for a transatlantic cable, but at the time it had seemed only a gentle swell of rising and falling terrain between the two continents. Voyages by the German research vessel *Meteor* in the mid-1920s and subsequent cruises by Woods Hole Oceanographic Institution ships provided hints that the ridge was more than it seemed. But the bathymetric profiles run by the *Vema,* and later by another Lamont-Doherty vessel, the *Conrad,* revealed that the ridge was in fact a vast undersea mountain range. Rising gradually from the depths, it was grooved by a valley deeper than the Grand Canyon. Bottom samples consisted of young, dark volcanic rocks, presumably from lavas hardening as they oozed through the crust into the cold water. The researchers also discovered that the Mohorovičić discontinuity, which marks the lower limit of the crust and is found at a depth of roughly twenty-five miles under the continents, was never more than about three miles below the ocean floor. At the bottom of the sea, only a thin membrane of rock lay between the ocean and the dense, hot material of the mantle.

Ewing asked graduate student Bruce Heezen to compile all the sounding data from the North Atlantic into a single map. With cartographer Marie Tharp, Heezen assembled a three-dimensional graphic rendering that showed how the central valley threaded its way the length of the North Atlantic's midocean ridge. Then, as they correlated this terrain with other geophysical

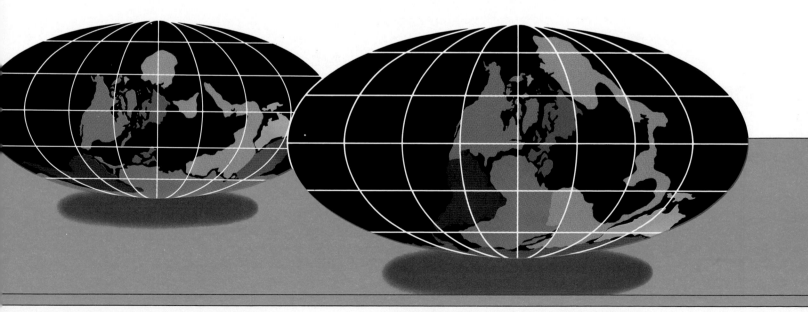

425 million years ago. Now sprawled over the South Pole, Gondwana moves north with a small T-shaped piece of land *(tan)* that will later collide with North America *(tan)* to become Florida.

306 million years ago. As Gondwana collides with North America *(tan)* and Europe *(purple),* sections of Asia *(light green)* rift away from Gondwana, heading northward.

data, Heezen and Tharp plotted earthquake epicenters on the map. The points fell in a band running right along the ridge. By matching other ocean ridges to zones of seismic activity, Ewing and Heezen predicted in 1956 that a rifted ridge would be found wherever the narrow band of earthquake epicenters could be traced through the centers of the oceans. Two years later, at Scripps Institution of Oceanography in La Jolla, California, Henry Menard confirmed that mountainous fissures almost precisely bisected the Atlantic, Indian, and Antarctic oceans. Like a giant seam in the planet's crust, the system ran for a total of some 34,000 miles, almost half again the Earth's circumference.

GEOPOETRY

The discovery of the oceanic ridge system seemed to raise more questions than it answered. Maurice Ewing himself did not know what to make of it. The youthfulness of the bottom samples at the rifts suggested that new seafloor was being created as material rose from the mantle. But although Ewing was willing to consider that this process added new material to the ocean bottom, he refused to believe it had any further implications: The expanding floor merely collided with the continents, and all motion stopped there. Others looked at the maps composed by Heezen and Tharp, however, and saw Wegener's old proposal for continental drift in a new light.

One scientist found the rift discovery enormously exciting. To Harry Hammond Hess, head of Princeton's geology department, the underwater rifts held out a possible explanation for something that had puzzled him for years. Twelve days Ewing's junior and another early convert to marine geology, Hess had joined the Naval Reserve before World War II in order to continue his oceanic studies. During the war, as captain of an attack transport in the Pacific, he often used an echo sounder to maneuver his ship in shallow island waters. Hess would keep his sonar running whenever under way, and by war's

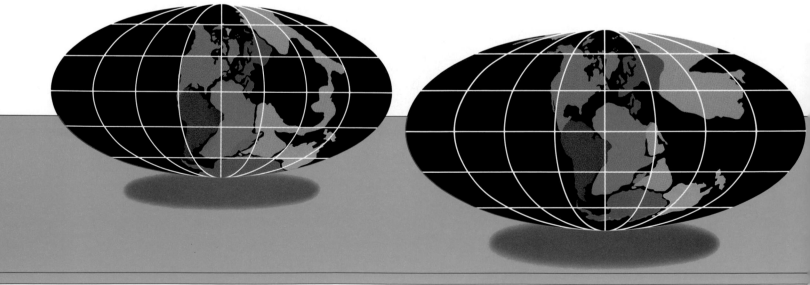

255 million years ago. The Americas *(tan and brown)*, Africa *(blue)*, Europe *(purple)*, Antarctica *(dark green)*, Australia *(pink)*, and much of Asia *(light green)* create the supercontinent Pangaea.

152 million years ago. The northern half of Pangaea breaks up. Fragments begin to form Europe *(purple)* and Asia *(light green)*; North America *(tan)* separates from Africa *(blue)*.

end he had mapped ocean depths along the hundreds of miles that his ship had traveled. Among the things he found in the Pacific were more than a hundred curious, flat-topped mountains rising thousands of feet from the seafloor. Hess called them guyots, after the Swiss American Arnold Guyot, Princeton's first geology professor. The isolated seamounts had evidently formed as volcanic islands along a midocean ridge, and their cones had been worn away by erosion in the atmosphere. Then, somehow, the blunted peaks had submerged. Hess later discovered that the farther the guyots were from the ridge, the deeper they were underwater, suggesting that they had somehow moved away from their birthplace.

The oceanic ridge system described by Ewing and Heezen, with its constant upwelling of molten material, appeared to offer a solution to the mystery of the guyots as well as the otherwise inexplicable youthfulness of the seafloor. Hess pondered the matter over the next several years, and in 1962, he presented his theory in a paper that began with the diffident disclaimer: "I shall consider this paper an essay in geopoetry."

Instead of continents plowing through a static ocean crust, as Wegener had proposed, Hess suggested that the ocean floors themselves were in motion, rather like conveyor belts driven by slow-moving currents in the underlying mantle. At the midocean ridges, rising mantle material released molten rock, which oozed out to form new ocean bottom along either side of the ridge. As new crust kept forming near the ridge, it pushed older sections away at the rate of about half an inch per year. In the process of spreading out from midocean, the seafloor shoved continents like Europe and North America apart. The one flaw in his conveyor belt scheme was that he lacked an explanation for where the descending end of the belt went. If Earth was not expanding to accommodate the new seafloor—and no one had offered evidence that it was—the seafloor had to be going some-

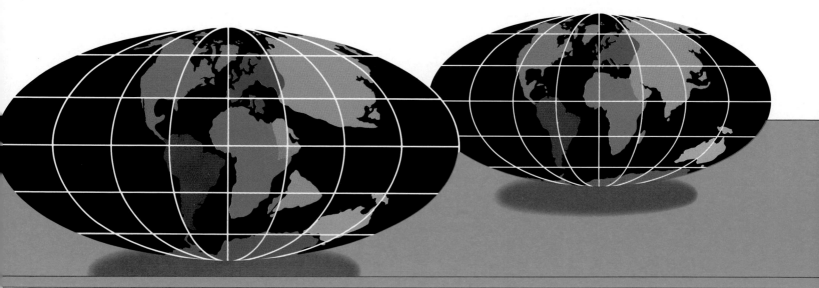

94 million years ago. Southern Pangaea splits into the puzzle pieces of South America *(brown)* and Africa *(blue)*. India *(light green)* goes north, Antarctica *(dark green)* and Australia *(pink)*, south.

50 million years ago. India *(light green)* collides with Asia, pushing up the Himalayas, and a landmass made up of Italy and Greece *(purple)* joins southern Europe, creating the Alps.

where. Perhaps, he proposed, it was taken in by the deep trenches that ring the Pacific. But without proof of some sort, the whole theory teetered.

MAGNETIC ZEBRAS

That proof was in the process of being gathered half a world away. In late 1962, aboard H.M.S. *Owen,* thirty-one-year-old Drummond Matthews was thinking about how to get as much geophysical data as he could from a six-month survey of the Carlsberg Ridge, which curves southeastward from the Gulf of Aden. A new Cambridge Ph.D., Matthews had been sent out to supervise the British research contribution to the International Indian Ocean Expedition. The *Owen* carried a magnetometer, an instrument that senses the strength and direction of Earth's magnetic field. Besides measuring the total present-day field, the instrument could also detect the almost imperceptible remnant of ancient magnetism. As magma wells up from beneath the crust and cools, it is imprinted with the strength and orientation of the magnetic field at the time, creating, in effect, a geomagnetic recording.

When the Earth's present-day magnetism was subtracted out of the ship's magnetometer record and the results plotted on a map, a pattern of magnetic variation emerged that was as distinctive as a tire tread. The record showed alternating intervals that seemed to correspond to periods of higher and lower magnetization. Shading in the strips of high magnetization produced a kind of zebra stripe, with alternating black and white wriggles that tended to parallel the ridge.

Back in Cambridge, Matthews pondered the mysterious data from the Indian Ocean with a new research student, who had the skills in geology, mathematics, and physics to appreciate both marine geology and the workings of Earth's magnetic field. Twenty-two-year-old Frederick Vine was also aware of Hess's ideas and much impressed by them. In late 1962, Vine began

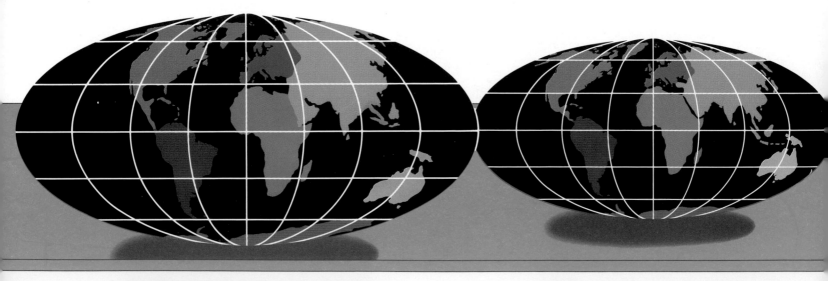

27 million years ago. Between Africa *(blue)* and Arabia *(light green),* the body of water now known as the Red Sea begins to open, as Asia and the Middle East continue to assemble.

Present. Antarctica covers the South Pole as Australia and Africa continue northward. The land bridge of Central America connects North and South America; the Asian continent is complete.

to look at the recordings in terms of seafloor spreading. The pattern seemed to indicate that ocean-floor rocks differed in their susceptibility to magnetization. Although on a much grander scale, the subtle variations in intensity measured at sea resembled those recorded on land, where rocks preserved evidence that the geomagnetic field had reversed itself in the past *(page 43)*. Sometimes the field was oriented as it is today, but sometimes it weakened and flopped, putting the present north magnetic pole in Antarctica and the south magnetic pole in the Arctic. As it happened, however, contemporary researchers working on the terrestrial magnetic record had only discussed their findings informally, and no one really knew what generated geomagnetism, much less how it could fluctuate.

But Vine made a great intuitive leap. Perhaps, he and Matthews theorized, the seabed had recorded the flip-flops of Earth's field as a series of strong and weak magnetizations as new molten rock came up from the mantle, cooled, and was gradually moved aside. A strongly magnetized band would form as long as Earth's field had one orientation, a weak one when the field was oriented the other way. Although the hypothesis was at this point pure speculation, Vine and Matthews submitted their paper to the British journal *Nature*, which published it in September 1963.

HELP FROM BELOW

The article attracted little support. Many scientists believed the Vine-Matthews theory did not connect to the geophysical reality of the ocean floor. If, as the two researchers proposed, the magnetic stripes represented sections that had been created sequentially, the stripes ought to be symmetrical on either side of the ridge crests, and there was as yet no clear evidence of that. To confuse the issue further, other oceangoing scientists were finding that the ridges were not smooth and unbroken but were instead cut at right angles by

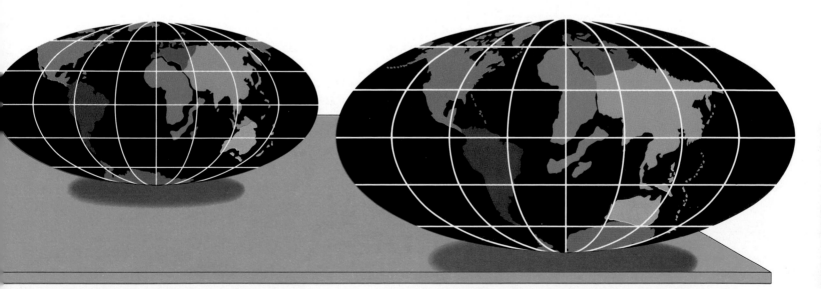

50 million years from now. According to geologic projections, Europe and Africa will converge, closing the Mediterranean; a strip of eastern Africa may detach along present-day rift lines.

150 million years from now. Africa will drift northward, as the Eurasian-African landmass rotates clockwise. Australia *(pink)* and Antarctica *(dark green)* will form a new southern continent.

great faults called fracture zones, creating a jagged, Z-shaped line. The straightforward movement of the theoretical conveyor belt envisioned by Harry Hess did not seem to fit the facts.

Answers came from seismology. In the 1960s, the United States Department of Defense fielded the Worldwide Standardized Seismograph Network, a $10 million application of state-of-the-art technology to the detection of nuclear tests. With 120 stations and rapid computer analysis, the instruments recorded seismic vibrations with unprecedented fidelity. And since nuclear detonations were, in effect, artificial earthquakes whose positions were known precisely, seismologists could track the behavior of seismic waves as they spread through the planet.

One of the first compilations of epicenter data from the new network was published in 1963 by Lynn Sykes, a young seismologist at Lamont-Doherty Geological Observatory, who plotted the location of earthquakes in the South Pacific. Seven years earlier, Bruce Heezen had been able to plot earthquake activity only as a poorly resolved swath along the midocean ridges. Sykes, in contrast, showed a pattern so well defined that epicenters could be correlated with specific topographic features. Earthquakes occurred along the ridges and also on the fracture zones that cut across them. But, as Sykes found, earthquakes happened along the fracture zones only between the two segments of ridge linked by fractures. Where these cracks in the seafloor extended beyond the ridges, they were, for some reason, seismically quiet.

AN INTUITIVE SOLUTION
The enigma was solved by J. Tuzo Wilson, a Canadian professor of geophysics then at the University of Toronto. The relationship between earthquakes and the seafloor had intrigued him for some time. A few years earlier, Wilson had been in Hawaii on his way home from the Antarctic. He had recently seen Hess's very preliminary thoughts on the seafloor as conveyor belt and was struck by the resemblance between the Hawaiian chain—a line of volcanic islands running from southeast to northwest—and Hess's description of the underwater mountain chains he had named guyots. Wilson decided the Hawaiian Islands must have been created as the seafloor crept over a stationary hot spot in the mantle.

Two years later, he applied this intuition to the patterns revealed by Sykes's article. Fiddling with a sheet of paper, he fashioned a rough model of the Mid-Atlantic Ridge, first making one vertical cut halfway down the sheet, and then another, parallel to the first but in the lower half of the sheet. Finally, he made a horizontal cut across the middle, between the bottom of the upper line and the top of the lower one. The result was a sheet severed top to bottom by a zigzag line. Fitting the pieces back together on a tabletop, Wilson then moved them apart. As the two halves were pulled away from each other, a widening patch of tabletop appeared between them.

This, Wilson saw, corresponded to new seafloor being created from below and spreading outward from the ridges. Along the horizontal cut, no tabletop

50

Deep beneath the oceans, molten rock from Earth's interior oozes from cracks in submarine ridges, adding new material to the vast moving plates made up of the lithosphere and the crust *(pages 56-57)*. As the new rock cools and hardens, it records the orientation of the planet's magnetic field. Over millions of years, the constant upwelling of fresh material pushes older seafloor away from the ridges, causing them to be offset by horizontal fractures in the crust.

The geophysical record reveals that the midocean ridges are symmetrically flanked by alternating bands of strong magnetization, corresponding to periods when Earth's field had its present orientation—with magnetic north at the North Pole—and weak magnetization, marking intervals when the field's polarity was reversed. If the strong bands are shaded in, as in the illustration above, the result is a zebra-striped pattern. Matching the stripes against magnetic reversals of radioactively dated rocks on land correlates different portions of the ocean bottom with successive geomagnetic epochs.

appeared as the halves were moved apart; no new seafloor was being created there. But because the two sides of the horizontal cut—the fracture zone—moved in opposite directions, they rubbed against one another and produced earthquakes. Beyond the segment of the fracture zone linking the ridges, however, there was no differential movement, or shearing, and so no earthquakes. Because the fracture zones marked areas where the outward movement of the seafloor from the ridge is transformed into a shearing movement along the fault, Wilson called this transform faulting. In 1965, a confluence of circumstance and common interests brought Wilson, Hess, and Vine together at Cambridge. Wilson arrived in January, followed soon after by Harry Hess, on sabbatical leave from Princeton. Most of the British geophysicists were away on a research cruise, but Fred Vine had stayed behind to complete his thesis. Working together for the first time, the trio began to hone their various observations and intuitions into a theory that would match the real world.

They began with California's San Andreas fault, which had the earmarks of an active transform fault; it seemed to vanish at its northern end, suggesting that it might connect to a ridge linked to another fracture zone. If Wilson's theory was correct, they would find a so-called spreading center—a ridge where new seafloor oozed out from below—between the northern terminus of the San Andreas and the southern end of the Queen Charlotte Islands fault farther north. Vine brought out a geophysical map made some years earlier for the ocean region off the American northwest, and the trio plotted earthquake epicenters and magnetic survey data on it. Sure enough, the epicenters ran along the faults, and magnetic stripes indicated that the seafloor had spread out from a feature they named the Juan de Fuca Ridge. In this case, the Z configuration lay on its side and had one very long bar—the San Andreas fault is more than six hundred miles long—and a very short stem, the ridge where spreading occurred. Here at last was solid evidence for a moving seafloor. Vine and Wilson published their Juan de Fuca paper in October 1965. In it, they proposed a model of seafloor spreading in the northeastern Pacific off Canada at an estimated rate of about three-quarters of an inch per year, marked by roughly symmetrical bands of reversed magnetism moving outward on both sides of the ridge.

There was one problem, however: The timing of known reversals in Earth's magnetic field did not quite match up to the stripes flanking the Juan de Fuca Ridge. But when Vine, who had joined Hess at Princeton, attended the November 1965 meeting of the Geological Society of America, he learned that another flip of the magnetic field had been found. Called the Jaramillo event, after the New Mexico creek near where the crucial samples had been collected, the reversal had occurred only 900,000 years ago, inserting a short

Monumental Collisions

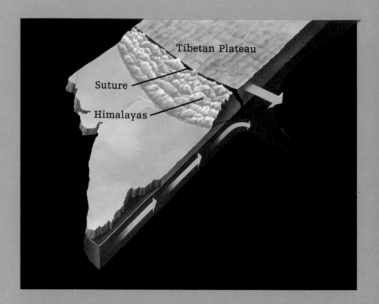

Seventy million years ago, India was an island continent, sailing northward through forgotten oceans aboard one of Earth's traveling plates. After a journey of 4,000 miles and 30 million years, the continent's inexorable progress brought it to the Eurasian coast. Slowly, and with devastating power, the huge landmass rammed the southern rim of what is now Tibet. The collision lifted the Tibetan plateau three miles above sea level, while at the same time crumpling the northern edge of India into the skyscraping peaks that are now known as the Himalayas.

This is the scenario currently favored by geophysicists to explain the origin of the world's highest mountains. A wealth of evidence supports the theory. For instance, along the probable line of collision, or suture, geologists have found rocks called ophiolites, which contain ocean sediments as well as material that could only have welled up from Earth's mantle. The scientists have also discovered sedimentary rocks in the northern Himalayas; not only do they add to the picture of Tibet as an ancient seacoast, but magnetic minerals within those rocks indicate that Tibet was shoved to its present location from more southerly latitudes *(page 43).* Moreover, the fossil record shows that mammals did not appear in India until after the presumed time of collision. These ancestors of today's goats and tapirs strongly resemble mammals once found in Mongolia, implying that the animals migrated down from China to populate the once-isolated land.

Geologists are not sure how the Tibetan plateau reached its present dizzying height, averaging 16,000 feet above sea level. The leading edge of India's plate may have slid beneath Eurasia, lifting up Tibet, or the collision may simply have smashed the area upward. The Himalayas themselves probably formed as the Eurasian plate bulldozed the Indian crust into overlapping slices; hence, the northernmost peaks of the Himalayas are the oldest.

The merger with Asia has slowed but by no means stopped India's northerly progress. Scientists estimate that the subcontinent continues to creep forward into China at a steady rate of about two inches per year, pushing that country and those of Southeast Asia north and east. Although the actual movement of the plate is virtually unnoticeable, its effects most certainly are not: Southern China suffers destructive earthquakes, including many that register as high as eight on the Richter scale.

period of normal polarity into an otherwise reversed epoch. The addition of a relatively recent change in magnetic direction smoothed the time scale Vine and his coworkers had used and evened out what had seemed to be an erratic rate of spreading at Juan de Fuca.

At about the time Vine was learning about the Jaramillo reversal, Walter Pitman III, a graduate student at Lamont-Doherty, was working up geomagnetic data from the *Eltanin,* a research vessel that had recently carried out a detailed survey across the Pacific-Antarctic ocean ridge east of Australia. The profiles from legs twenty and twenty-one of the *Eltanin*'s crisscrossing course resembled the profiles for the Juan de Fuca Ridge. Leg nineteen was something else again. It "hit me like a hammer," Pitman remembered later. The data looked "the way a profile ought to look and never does." The symmetry of the zebra pattern was a virtual set of mirror images on either side of the ridge. Although one of his colleagues—profoundly skeptical of the moving seafloor theory—thought the data "too perfect" and thus knocked "the seafloor spreading nonsense into a cocked hat," Pitman was convinced.

Supporting evidence continued to accumulate. In late 1965, researchers at Lamont-Doherty who were examining sediment cores taken in the southern Indian Ocean by the *Vema* five years earlier found a 900,000-year-old reversal event that matched the *Eltanin* magnetic profiles. And in 1968, the drill ship *Glomar Challenger,* upon punching holes into crust on either side of the Mid-Atlantic Ridge, revealed that the ocean floor grew older with increasing distance from the ridge, just as theory predicted.

THE DEEP

Once members of the geologic community fully accepted the idea that the seafloor was continually renewed from fissures in midocean ridges, they looked with a fresh eye on some old observations and hypotheses. As early as the 1930s and 1940s, American and Japanese seismologists had been puzzled by earthquakes that seemed to originate at depths of more than 400 miles. Most earthquakes in their experience occurred in the shallow layers of the crust; at the time, theory held that material below the crust was too fluid to snap and cause earthquakes. More puzzling still was that these deep-focus quakes, as they are known, tended to happen on a plane that inclined steeply downward from offshore trenches toward the continents of Asia and South America: Quakes near the trenches were relatively shallow; the closer they were to the continents, the deeper their source.

In 1954, Hugo Benioff, one of the world's leading designers of seismic instrumentation, proposed that deep-focus tremors occurred within great sheets of ocean floor that were being thrust down into the mantle. Nearly a decade later, Harry Hess suggested this same mechanism to account for what happened to the seafloor as it moved out from midocean. Benioff lacked a theory to explain what was driving the ocean floor; Hess had the theory but lacked supporting evidence for it. Once that evidence grew weighty enough, however, subduction, as the process is called, was recognized as the missing half of Hess's conveyor belt. Where the thin layer of expanding seafloor collides with the thicker continental crust, it is forced down and ultimately back into the internal crucible where it formed eons earlier. The deep trenches mark the site of this massive encounter.

By the mid-1980s, geophysicists better understood the relationship among the many parts of Earth's architecture. It was an accepted fact that the planet's surface is composed of huge, rocky plates whose movement rips continents apart and shoves them together, stretching and folding the face of the land and ocean floor over vast reaches of time. But the internal dynamics remained largely a mystery. No one knew the details of what happened at depths greater than about 400 miles, where earthquakes ceased, or how far into the Earth's interior the plate-driving mechanism extended.

To peer into the hidden heart of the planet, researchers at Harvard and Caltech invented seismic tomography. Tomography, a term derived from the Greek word for "slice," effectively takes a kind of sectional x-ray through the Earth, using differences in seismic-wave velocities to distinguish contrasting regions in the mantle. Regions with high seismic velocities are generally taken to be composed of colder, more rigid material; those with low velocities are assumed to be made up of very hot, more pliable material. A large computer builds the two-dimensional slices into a three-dimensional view.

THE ANTIWORLD

Up to a point, the snapshots have reinforced prevailing views of the interior. To a depth of 100 or so miles under the midocean ridges, large regions register the slow seismic velocities that correspond to molten rock rising from the mantle. And beneath continents, where subducted plates dive back into the mantle, high velocities indicate relatively cold material to at least 125 miles. But below about 200 miles, things begin to lose their definition. The subtle distinctions between seismic velocities become increasingly difficult to discern even with state-of-the-art instrumentation, causing the boundaries between regions of different temperatures to blur. Seismologists see something, but there is no consensus on what the shadowy shapes mean.

Some scientists believe, for example, that they have detected the deep roots of Africa and South America where the two continents appear still to be joined. Other researchers, eyeing readings at the 400-mile depth, say that the mantle shades here into a zone of much denser material, in which enormous pressures cause minerals to change their crystal structure, forming a natural barrier that prevents the upper and lower regions of the mantle from mixing.

Still others perceive a kind of inside-out antiworld at the boundary between the mantle and the planet's fluid outer core. As huge slabs of cool material— the sinking wreckage of subducted ocean floor—descend through the mantle, some researchers hypothesize, the fragments pile up along this boundary, forming a superdense anticrust. From this crust, undulations hundreds of feet high thrust down into the liquid outer core, where temperatures reach more than 6,000 degrees Fahrenheit, comparable to the surface temperature of the Sun, and pressures are more than a million times that at the Earth's surface.

Hundreds of miles deeper into the planet's center, temperatures rise to some 9,000 degrees and pressures more than double. Within the confines of the inner core, they are so great that even the tremendous heat at the center of the Earth cannot melt iron. In ways not clearly understood, the interactions between these two inhospitable worlds, one molten metal, one densely solid, began to spin Earth's web of magnetism at least 3.5 billion years ago and have been at it ever since. Occasionally, the mechanism may have faltered, causing the field to flicker and reverse—and create the surface effects that ultimately verified plate tectonic theory. No doubt the secrets of the central dynamo will someday be discovered, but for the moment the realm 2,000 miles below our feet is as mysterious and alien as anything to be found among the stars.

RESTLESS PLANET

Beneath its delicately toned skin, Earth seethes with a kind of slow violence that endlessly remakes the planet. Driven by superheated currents of molten rock coursing through the planet's interior, more than a dozen major and minor tectonic plates—multibillion-ton slabs of the Earth's surface, some so huge that they encompass individual continents—crawl across the face of the globe to inevitable collisions.

The motion of the planet's face is a never-ending cycle of geologic creation and destruction. Traveling only inches per year, some giant fragments of the planet's fractured surface collide, while others pull apart with a force that can rip whole continents asunder. Where plates separate, magma may well up from the planet's interior to fill the gaps with new rocky material. Where plates clash, one edge is forced down into the hot rock of the mantle, a process that scars the ocean floor with deep trenches.

As shown on the following pages, the Earth's most dramatic features are a consequence of the interactions between the planet's perpetually shifting surface and its interior inferno. But not all of the violence is of a gradual kind. Along the boundaries between the plates and over superheated plumes of magma, periodic earthquakes, volcanic eruptions, and other geologic cataclysms give convulsive testimony to the torments of the world below.

LAYERS OF
HEAT AND PRESSURE

The Earth's tectonic plates form the outer layer of a dynamic structure whose temperatures and pressures are so high that rock flows weirdly through rock. The planet's core, a dense ball composed primarily of nickel and iron, has two layers. Innermost is a sphere some 1,450 miles in diameter, squeezed so tightly by the mass of the planet that its metallic elements remain in solid form despite heat of more than 9,000 degrees Fahrenheit, far in excess of their surface melting points. Wrapped around this solid

Nine major plates and a number of minor ones jostle and shove their mighty way around Earth's fractured surface, carrying the continents with them. The plates are slabs of lithosphere, propelled by heat from the soft asthenosphere below. Some—such as those under the Atlantic—are pulling apart, moving the Americas farther from Africa every day. Others, such as the one carrying India, have been involved in gigantic collisions within the past 40 million years.

core is an outer core of molten metal 1,430 miles thick.

Between the core and the crust lies the restless, three-layered mantle. Consisting mainly of lighter elements such as silicon and oxygen, the mantle is considerably less dense than the planet's core. Scientists know little about its lowest layer, the mesosphere (orange), which measures about 1,550 miles from top to bottom. The asthenosphere (dark orange), a middle section 120 to 250 miles thick, is so hot that its rocky material would flow like liquid were it not for the intense pressure that holds it in a plastic, semisolid state. Uppermost is the lithosphere (red), a cooler, more rigid husk of solid rock up to 100 miles thick. Heat from below has deformed the brittle lithosphere, breaking it into plates that float on the asthenosphere

like shell fragments on the surface of a massive egg.

Harbored within the uppermost lithosphere is the Earth's crust (brown), made up of the continents and ocean floors. Containing a variety of light elements, including silicon, oxygen, aluminum, potassium, and sodium, the crust rests within the lithosphere, and the lithosphere in turn rides on the softer asthenosphere. At its thinnest—beneath the oceans—the crust is as shallow as a few miles deep. Where continents rise above the oceans, it can be thicker than thirty miles. Like vast icebergs with most of their substance submerged, continental mountain ranges extend deeply into the soft underlying layer, so that their invisible bases serve as stabilizing keels beneath the floating continental masses.

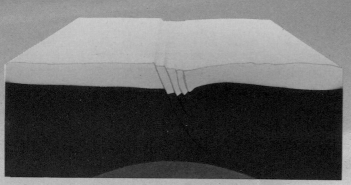

In the early stage of rifting, shown in the southern part of a rift running from Mozambique to Ethiopia, rising hot mantle rock stretches the lithosphere, opening a crack in the crust.

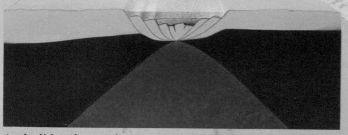

As the lithosphere and crust stretch over the heated plume of rock, an ocean basin begins to form. Water may flow over the continental floor, as it does in the rift's western branch.

In the southern Red Sea, rifting is more advanced. The lithosphere has cracked, and basaltic magma wells through to create a narrow swath of new oceanic crust, spreading the seafloor.

In the Gulf of Aden, located to the south and east of the Red Sea, the ocean basin widens as the spreading seafloor slowly separates the two landmasses.

FORCES THAT TEAR THE LAND ASUNDER

The stirrings of rock within the Earth's mantle have ripped and reshaped the world's surface for at least a billion years and perhaps many more. Stretched by heat from below, plates carrying the planet's crust have repeatedly split apart, leaving scars down the middle of the oceans and around the rims of many continents. At times, great landmasses have torn in two, as is visible today in eastern Africa *(right)*.

Evidence of rifting, as this action is called, can be found around the globe in the form of oceanic ridges marking the boundaries between separated plates. These rocky crests appear when magma, molten rock from the mantle, wells up through the rift and creates new oceanic crust. In some places, volcanic peaks rise along the separation zone and appear above the waves as islands.

When continents pull apart, seawater may eventually rush in to fill the space between the separating landmasses. Meanwhile, the rift widens and new seafloor grows under the water. Four stages of this process can be seen within two parts of the East African rift, as well as the seafloor beneath the southern Red Sea and the oceanic basin in the Gulf of Aden *(left)*.

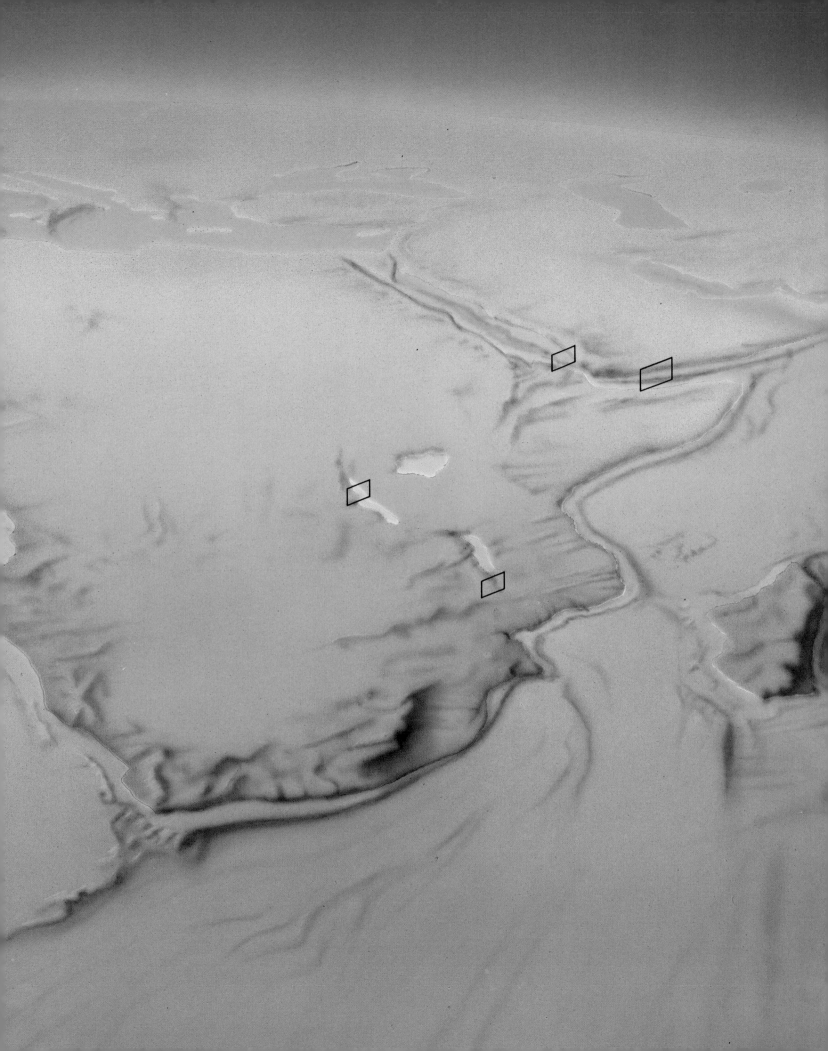

As the South American Plate bears down on the Nazca Plate, it presses a great slab of oceanic lithosphere deep into the asthenosphere, creating the Peru-Chile Trench. Melting rock from the leading edge of the Nazca Plate, mixed together with dissolved gases, erupts explosively through Andean volcanoes.

THE RECYCLED EARTH

Even as some plates are split or pulled apart to create new oceanic crust, others are slammed together to destroy vast areas of the planet's surface in a geologic process known as subduction. When one drifting oceanic plate smashes into another, or into a continental plate, one of them bends sharply downward and plunges into the asthenosphere, where the planet's raging interior furnace melts and recycles some of its rocky material.

During a collision, the subducted plate can drag the edge of the other plate with it, denting the Earth's surface with a trench where the two rock masses meet; in the western Pacific's Mariana Trench, for instance, the subduction of the Pacific Plate has produced a trough reaching to 36,000 feet below sea level—deep enough to swallow Mount Everest. The force of the collision usually causes the overriding plate to buckle, and melting of the subducted plate produces an arc-shaped chain of volcanic mountains, such as those in the western Pacific.

The subduction process can also be seen vividly along the Pacific coast of South America, where the westward-moving South American Plate pushes the eastward-moving Nazca Plate deep into the mantle. The entire Andean coast is regularly subjected to severe earthquakes, a consequence of the enormous stresses associated with the movement of the subducted plate. Adding to the spasmodic violence are devastating volcanic eruptions, which occur when buoyant magma from the melting plate rises through the overriding crust.

TRANSFORM FAULTS: EARTHQUAKE MAKERS

As the Earth's tectonic plates swim slowly around the globe, some neither collide nor pull apart but instead slide along a fault zone. The friction of two great slabs of rock shouldering past one another along a transform fault zone, as this area is called, may produce thousands of earth tremors per year, ranging from barely detectable shocks to calamitous earthquakes.

Transform faults in the oceanic crust tend to run for relatively short distances, joining segments of oceanic ridge to form zigzag patterns; other such faults may dissect a portion of a continent. One of the world's best-known continental transforms is the San Andreas fault, part of an extended and active system that runs for more than 600 miles from the waters off northern California into the seafloor near Mexico's Pacific coast. Along this fault zone, the Pacific Plate, carrying a small slice of continental lithosphere on its eastern edge, rubs northwestward against the western edge of the North American Plate.

Moving at a combined separation rate of about two and a half inches per year, the plates slide smoothly along their opposing courses in some places. Often, however, their irregular edges snag, and stress builds as the two plates remain locked together. Eventually the stress reaches the rupture point, and portions of the plates lurch suddenly, unleashing shock waves capable of producing damage hundreds of miles away. The great San Francisco earthquake of 1906, which measured about 8.3 on the Richter scale, was probably occasioned by just such a buildup and release of stress along a lengthy segment of the fault.

When two plates slide past each other, their mutual boundary is called a transform fault—so named because the fault ends when it is transformed into either a spreading ridge, where the plates begin to separate, or a subduction zone, where they begin to collide. When the plate edges catch and then suddenly release, the earth jerks forward, generating an earthquake.

THE BIRTH OF VOLCANIC ISLANDS

Although tectonic drama usually occurs where segments of Earth's surface collide, tear apart, or slide past each other, violence is not confined to the edges of plates. Some of the world's largest volcanoes, for instance, are found far from the boundaries of the plates on which they ride. The Canary Islands in the Atlantic and several island chains in the Pacific are the summits of gigantic undersea volcanoes that formed within plates. So too are the Hawaiian Islands, situated at the southeastern end of a chain of active and extinct volcanoes—many now completely submerged and eroded by the sea—stretching across the northern half of the massive Pacific Plate.

Geologists believe that volcanic islands of this type bubble up over so-called hot spots underneath the moving plates. According to theory, the temperature at certain points deep in the Earth's mantle is unusually high. From these hot spots, long plumes of molten rock rise, burn through the lithosphere, and spill out as magma to build conical volcanic mountains. The Earth's interior may hold twenty or more of these spots; any one of them can give birth to a long chain of volcanoes as a plate moves over the relatively stationary heat source. The island of Hawaii, the youngest and most volcanically active member of its chain, is currently closest to the hot spot that has been spewing up the stuff of new volcanoes as fast as the Pacific Plate can carry the old ones away to the northwest. At the edge of the plate, up near the Aleutian Islands, are extinct volcanoes that once erupted in Hawaii's place.

Theoretically, many midplate chains of volcanoes are born when a hot spot deep inside the Earth's mantle burns through the moving plate above it. The plate gradually moves the volcano off the heat source, and it becomes dormant even as a new peak forms beside it. In the last 65 million years, the spot below the Hawaiian Islands has created a line of volcanoes 4,000 miles long.

Winds born of the exchange of heat between sea and air whip moisture laden clouds into a cyclonic storm over the Pacific Ocean.

The morning's patch of unsettled weather moved swiftly out of Africa's eastern desert, carried on a deep atmospheric current running counter to the planet's spin for more than 20,000 miles. Satellites detected its telltale cape of storm clouds and flashed alerting images to scientists on the west coast of the continent. Within hours, a multinational squadron was flying seaward from its base, poised for interception. Six hundred miles southwest of Senegal, the first in an ocean-spanning armada of research ships took up positions in a giant hexagonal array. Under the blazing tropical Sun, their crews readied equipment that would dissect this latest whorl of weather from its cloud tops to its energy source in the equatorial water below.

It was a familiar routine. In the summer of 1974, some 4,000 scientists and technicians from seventy-five nations had converged on the Senegalese capital of Dakar, a modern port city of white buildings and red tiled roofs. With the army of people came thirty-eight research ships, thirteen aircraft, and a varied collection of data-sensing ocean buoys. Day after day for a period of three months, they monitored the intermittent advances of meteorological disturbances, hoping to answer fundamental questions about the coupling of atmosphere and ocean.

The research program was the first major attempt to comprehend a large segment of the planet's two fluid covers, the dense and sluggish liquid ocean of water and the thin and volatile gaseous one of air. The project's sensors were concentrated between the continents of Africa and South America and pulled in data from a mile beneath the sea surface to a height of twenty miles in the atmosphere above. But the study had a broader focus, taking in a volume of equatorial atmosphere about twenty degrees of latitude wide and covering 20 million square miles from Ethiopia around the world to the Indian Ocean.

Along this band just north of the equator, energy from the Sun is stored in the warm ocean, then carried by water molecules into the atmosphere by wave action and evaporation, beginning a long journey toward the polar ice fields *(pages 90-91)*. The movement of heat energy from the equator, which has too much, to the poles, which have too little, smooths the distribution of solar warmth. Over short intervals of time, the atmospheric cargo of waterbound energy generates the ephemeral phenomena of weather. Over a longer term, it moderates and shapes terrestrial climate.

The process begins with interactions between the great westward flow of the ocean currents and the prevailing trade winds, the easterlies blowing from Africa to the New World and across the Pacific north and south of the equator. The currents of the sea mirror the circulation of winds and are largely driven by strong variations in temperature within the atmosphere and ocean. From the warm ocean near the equator, the currents carry heat on huge streams that loop toward higher latitudes, where they cool, release their burden of energy, and turn back toward the equator for more. About 40 percent of the heat moved from the equatorial sea goes by water; the rest travels in the atmosphere.

To the planners of the 1974 project, the transport mechanism itself appeared to intensify in what is called an easterly wave—a U-shaped bend in the westward blowing winds that moves like a kink in a rope, propagating out of Ethiopia's desert mountains to circle the globe, sowing heavy weather as it goes. "The waves were always a puzzle," explained Joachim Kuettner, the German-born international director of the study, remembering the project fifteen years later. "We wanted to know how they existed, where they took their energy from. They produce enormous, intensive squall lines just south of the Sahara along the west coast of Africa. Over the water, there are tremendous rains and sometimes hurricanes. Traveling ten or twelve degrees of longitude a day, the atmospheric waves move into the Caribbean and Gulf of Mexico, cross Central America, and continue across the Pacific until they are stopped in the Indian Ocean by the monsoon, which is a very strong system. You don't see the waves themselves. But you can see the weather they are producing."

In 7,000 computer tapes of data extracted from that long, hot summer off Africa, scientists acquired their most detailed picture to date of how the air and ocean interact in the tropics. They found that heat is sent skyward by explosive surges of vertical motion, or convection, in the weather systems of the easterly wave's U. What had been a blur on the blueprint of air and ocean sharpened into a mechanical detail that, a few years down the road, would be put to work in the day-to-day routines of tropical meteorology and hurricane forecasting.

In the wake of the tropical project, the United Nations' World Meteorological Organization in Geneva and the Paris-based International Council of Scientific Unions collaborated on a series of ambitious follow-up studies. One program, begun in 1978, provided a full year of intensive, worldwide observations of the atmosphere and ocean. A second project, also carried out in 1978 and 1979, explored the monsoon, the seasonal give and take of moisture between the Indian Ocean and subcontinent. A third, begun in 1985 and scheduled to run for ten years, is quantifying the linkage between air-sea interactions in the tropical Pacific and global circulations in the atmosphere and ocean. Taken together, such studies make up a kind of mosaic in which the key working parts of the air-sea system can be discerned.

All of these experiments reflect a growing recognition that, although the medium of water and the mantle of air seem as different as fish and birds, they are in fact two elements of a single system. Such studies have greatly clarified how weather is generated in the atmosphere and have described the slower adjustments that constitute the weather of the sea. Most important, they have helped scientists understand how these delicately balanced processes change over very long intervals of time.

RIVERS IN THE SEA

The seeds of these expansive efforts were sown centuries ago in the frustrations of early researchers, who had no way of acquiring data describing conditions all across the planet at any given time. The fundamental problem may have been best stated by John Rennell, a British Army major who had turned in retirement to studying the currents of the seas. "The want of simultaneous observations," he wrote in 1776, "is an incurable defect. By this we are kept in ignorance of the state of things in every other quarter, save the one in which our own observation was made."

Some of the ocean's large dynamic features were known in Rennell's time. In the years before the American Revolution, while serving as deputy postmaster general for His Majesty's colonies, Benjamin Franklin learned that British mail packets bound for the New World took some two weeks longer to cross the Atlantic than merchant ships captained by Americans. Puzzled, Franklin consulted his cousin, Timothy Folger, a Nantucket whaling skipper, who had a ready answer. In their pursuit of whales in the Atlantic, Folger and his colleagues had discovered a strong oceanic current, which they called the Gulf Stream because it flowed north out of the Gulf of Mexico. At the latitude of the Carolinas, it veered east toward England and Northern Europe. Because whales preferred to travel at the edges of the warm water, the hunters were obliged to cross and recross the Gulf Stream, gaining a detailed knowledge of it in the process. Westward-bound American captains thus learned not to sail against this powerful current, while the British mail packets resolutely plowed straight into it, costing themselves as much as seventy miles a day. Their captains, Folger observed, "were too wise to be counselled by simple American fishermen."

Folger sketched the Gulf Stream for Franklin, who had it added to an existing chart of the Atlantic. The masters of British packets spurned the information, which was not a bad thing, from Franklin's point of view, since their delays in crossing the Atlantic later worked to the advantage of the rebellious colonists. It was left to Franklin's great-grandson to make the first serious, scientific studies of the current. Beginning in 1843, Alexander Dallas

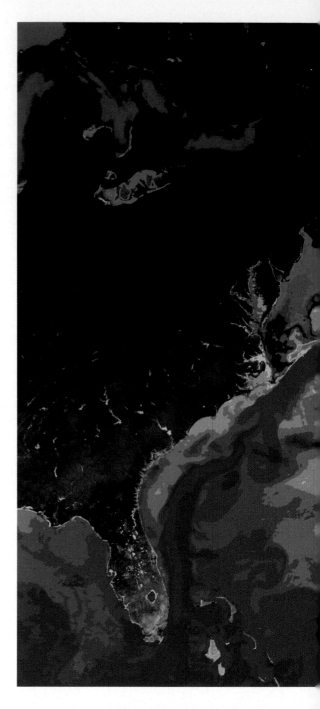

Carrying a hundred times the water flow of all the world's rivers combined, the Gulf Stream—shown in red, orange, and yellow in this heat-sensing satellite image—shuttles equatorial heat to the North Atlantic, tempering the icy Labrador Current (purple, pink) and spawning westerly winds that warm the British Isles. Eddies on the Gulf Stream's edge encircle cooler mid-Atlantic waters (green), trapping marine life in a ring dance that can last for months.

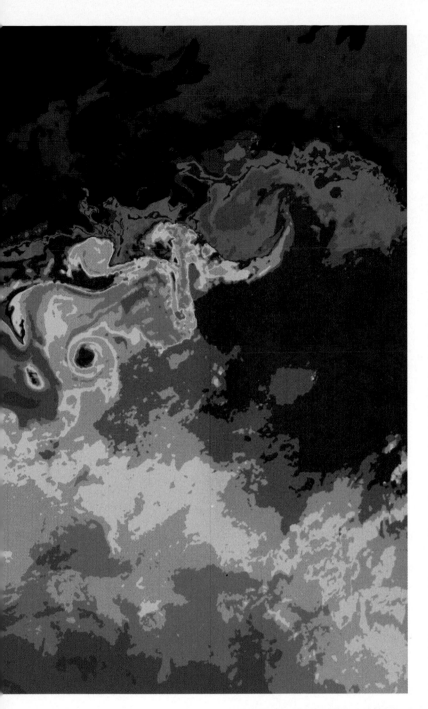

Bache, superintendent of the U.S. Coast Survey, ordered his ships to map the Gulf Stream and measure the temperature and depth of these warm, swift-flowing waters.

Scientific discovery then, as now, was often hotly competitive, especially when different government services were involved. Bache had an archrival, the U.S. Navy, as personified by Lieutenant Matthew Fontaine Maury, head of the Depot of Charts and Instruments. While Bache's vessels approached the Gulf Stream as part of their coastal mission, Maury dispatched a high-seas research vessel of his own—and outdid Bache by enlisting the cooperation of merchant sea captains in collecting data. In 1847, Maury published the data in the first of a series of wind and current charts, which were soon adopted by American sailors seeking a path of least resistance across the Atlantic.

Despite a promising start, American oceanographic efforts languished in the second half of the century, handicapped by a lack of funds for what one member of Congress derided as "this thing called Science." Interest in the oceans was on the rise in Europe, however. Prince Albert I of Monaco, for example, used some of his immense fortune to study the Gulf Stream by the simple technique of putting notes in bottles and tossing them overboard as he cruised about on his yacht. From responses to his hundreds of multilingual missives, Albert discovered that the Gulf Stream actually forks in midocean, with part of it making a clockwise turn to the south past Spain and Africa before heading west, back across the Atlantic toward America.

Hit-or-miss efforts like Albert's could do little more than sketch such great rivers in the sea. Yet major expeditions did not do much better. Between 1872 and 1876, the British research vessel *Challenger* circumnavigated the globe, its scientists collecting thousands of specimens of never-before-seen marine organisms, along with samples of seawater and mud from the ocean floor. Valuable as it was, the data existed in isolation; there was no overall theoretical framework that would integrate the *Challenger* observations. Scientists' understanding of the oceans still suffered from their inability to visualize the whole system.

Photographed from the *Apollo 6* space capsule, cloud formations over the Pacific mirror the knife-edge boundary between cold and warm ocean currents 120 nautical miles below. Warm waters give rise to buoyant air that builds closely bunched stratocumulus clouds *(far left)*. Looser, more stratified clouds *(left)* occur in the heavier, more stable air mass over a cooler sea.

The lack of simultaneous observations was an even bigger handicap to studies of Earth's atmosphere. Predicting the weather has always been a matter of great practical importance, but before the development of modern communications there was simply no way to use what little information was available; storms traveled faster than news about them. By the late nineteenth century, communications had improved enough to allow short-term forecasts based on maps showing "elements of present weather" in various locations. New maps were compared with their predecessors to see how the atmosphere had changed, and predictions assumed that what had happened before would happen again—that, for example, a pattern of temperature, humidity, and pressure that produced showers on one occasion would do so on the next.

During World War I, such maps were regarded as military secrets by the belligerent nations, and so were unavailable in neutral Norway. But meteorological pioneer Vilhelm Bjerknes and his son Jakob were already at work on an alternative scheme of weather forecasting. As a professor at the University of Bergen, the elder Bjerknes theorized that atmospheric behavior could be explained in thermodynamic terms. According to Bjerknes, the parade of Scandinavian weather came in cyclones, large spirals of ascending air created where warm and cold air masses collided in the atmosphere. Adopting the military jargon of the day, he described these boundaries as fronts, the term still used on weather maps. When a mass of cold air blew out of the polar regions and met a mass of warm air moving toward the poles from the tropics, the result was a disturbed area where warm, wet surface winds converged and rose, lowering atmospheric pressure. A cyclonic wind system blew counter-clockwise around this area of low pressure, pushing masses of warm and cold air into conflict along fronts extending from the center. Storms, like battles, broke out along such fronts and lasted several days.

Vilhelm Bjerknes concluded that "the rational solution of forecasting problems" required detailed data on global atmospheric conditions as well as "a sufficiently accurate knowledge of the laws according to which one state of the atmosphere develops from another." His son would pursue this work in America and ultimately develop a broad view of linkages between the atmosphere and the sea.

FORECASTING BY THE NUMBERS
Meanwhile, the challenge of solving the problems set by Vilhelm Bjerknes was taken up by a young, mathematically inclined British physicist, Lewis F. Richardson, while he was serving as an ambulance driver in France. The atmospheric behavior described by Bjerknes, Richardson believed, might be presented quantitatively, using numerical values for temperature, atmospheric pressure, and wind velocities at each of a number of locations. If such values were then plugged into mathematical formulas representing physical laws, he could produce accurate forecasts of the next day's conditions. Contrasting his approach to the air-mass studies of Bjerknes, Richardson wrote, "the arithmetical procedure is the more exact and the more powerful in coping with otherwise awkward equations." As a test, Richardson tried to predict the weather at a single spot—Hamburg, Germany—one day into the future, furiously calculating the necessary numbers during pauses between wartime ambulance missions. But the atmosphere moved with greater complexity and speed than Richardson could set down in numerical form, and his predictions proved utterly wrong.

Undaunted, Richardson dreamed of a day when his methods would be applied in a vast "forecasting factory" where platoons of human computers, the only kind then in existence, would process global data and solve the equations that had overwhelmed his one-man effort. Richardson may have failed to predict the next day's weather, but he had glimpsed the future.

War gave more to the study of atmosphere and ocean than Bjerknes's fronts and Richardson's imagined forecast factory. Battles were being waged both on and under the ocean, and airplanes had taken combat into the sky. After the Second World War, meteorologists used weather-adapted radars stripped from mothballed bombers to "see" water droplets in the atmosphere and apply what was seen to tornado and hurricane warnings. They also employed balloon-borne sensors that could trace vertical profiles of atmospheric temperature, humidity, and winds up to 100,000 feet, radioing measurements as they rose through the air. Radio and teletypewriter networks linked weather forecasters across the continents.

During the 1950s and 1960s, digital computers and Earth-orbiting satellites went to work for meteorology, supplying great quantities of information on global atmospheric conditions and methods for handling it. Weather watching became an international undertaking that would pave the way for widespread technical cooperation between nations.

Important as the day-to-day variations called weather were to commerce and war, they were trivial compared to the longer historical sweep of climate, which is the natural history of the atmosphere. Today's cloudy sky is no more relevant to climate than a traffic accident is to a civilization's rise and fall. Until the 1970s, however, the study of climate was regarded as somehow less critical than weather prediction, a coarse technique for deriving almanacs and trends of heat and cold, moisture and drought. Then climatology began to tell a disturbing tale of change, hinting at a world that might not be eternally habitable, but was instead delicately poised between the climatic abysses of sweltering Venus and frozen Mars—a world whose fragile balance could be tipped by nature or by humankind.

On a planet whose present atmosphere has been in place for some two billion years, a few centuries of recorded weather are slim pickings for anyone trying to guess the distant future. Fortunately, climate writes its histories in other languages as well—in the annual rings of growing trees, layers of polar ice and seafloor sediments, rock strata, fossils, and ancient chemistries. Such windows on the past reveal that the planet has sometimes been much warmer than it is today, and sometimes much colder.

Earlier in this century, an obscure Yugoslav mathematician named Milutin Milankovitch suggested why the cold phases might have occurred. In 1920, he proposed that slight variations in the planet's rotational axis and orbit periodically altered the distribution of solar energy on the Earth's surface *(pages 76-77)*. Although the changes were minuscule in absolute terms, they could cool northern summers enough that winter snowpacks endured and spread from year to year. As the amount of ice and snow increased, the bright surface reflected more and more solar energy back into space, thereby causing further cooling, more ice, and finally, the chilling periods now called ice ages. According to Milankovitch, these per-

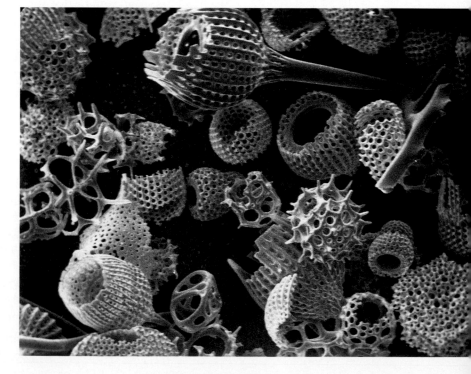

The lacy fossils of tiny planktonic creatures called radiolaria, dredged up in core samples of ancient seabeds and shown here in an electron micrograph, allow scientists to reconstruct hundreds of thousands of years of climate history. Certain species of radiolaria thrive only in cool seas, others only in warm. By counting the number of each species within dated layers of sediment and then analyzing the data with powerful computers, climatologists can determine fluctuations in average sea-surface temperatures.

turbations led to three distinct cycles of glaciation, at intervals of about 100,000, 41,000, and 22,000 years. Most of his scientific contemporaries were skeptical, however.

Previous evidence of such cycles, taken from sediments and studies of alpine glaciers, did not quite square with Milankovitch's astronomical theory. But in the 1970s, a more coherent record of these frozen interludes began to emerge from the tropical ocean. Because corals grow only near the ocean surface, they preserve a record of rises and falls in sea level in the form of a series of terraces. The lower terraces indicate cold periods when huge amounts of water were locked up in the polar icecaps, lowering the sea level. When the coral markers were radioactively dated in Pacific atolls, the Bahamas, and the Florida Keys, their history marched in step with the three periods inferred by Milankovitch.

MORE SUPPORT

Another biological record was stored in the extensive collection of seafloor samples housed at Columbia University's Lamont-Doherty Geological Observatory. These cylindrical cores, extruded by hollow probes that had been dropped like weighty darts into the ocean bottom, contained more than mud and rock; tiny animals had been trapped in layers deposited over millions of years. To a pair of inquiring scientists, James D. Hays of Columbia and Nicholas Shackleton of Cambridge University, the microscopic remains were a record of global climate.

Hays studied the skeletons of radiolarians, protozoans with elaborately filigreed external skeletons made of silicon compounds. The single-celled organisms occur in several species, which vary in population with changing water temperatures. By sorting out the proportions of each species found in the Lamont-Doherty core samples, Hays was able to estimate the average temperature of the seawater at the time the sample had been deposited. This permitted him to track climatic change over the course of hundreds of thousands of years.

Shackleton focused on the shells of a microorganism called foraminifera, which appeared to be a critical player in the vast chemical plant of the sea. The shells of sea creatures are formed as the animals extract the necessary chemicals from the seawater around them—mainly carbon, calcium, and oxygen. Shackleton was most interested in the oxygen content of foraminifera shells, since it was possible to match that against the occurrence of glaciation in the world. Two variants, or isotopes, of oxygen—oxygen 16 and oxygen 18—are found in the shells. The proportions of the isotopes varied, depending on how much of each was available in the ocean when the creatures lived. This, in turn, was determined by global temperature: Water molecules containing oxygen 18 are heavier than those containing oxygen 16 and hence are less likely to evaporate—especially at times of low temperatures. Thus, foraminifera shells high in oxygen 18 and low in oxygen 16 signal periods when an ice age was in progress.

CYCLES THAT TRIGGER AN ICE AGE

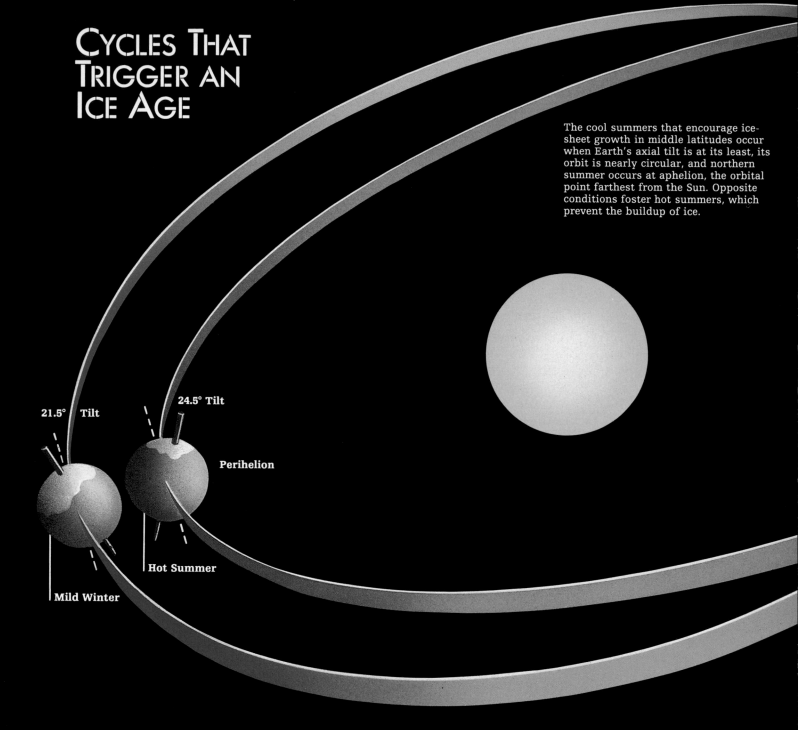

The cool summers that encourage ice-sheet growth in middle latitudes occur when Earth's axial tilt is at its least, its orbit is nearly circular, and northern summer occurs at aphelion, the orbital point farthest from the Sun. Opposite conditions foster hot summers, which prevent the buildup of ice.

21.5° Tilt

24.5° Tilt

Perihelion

Hot Summer

Mild Winter

At intermittent periods over the last million years, a persistent chill has gripped Earth. Less than 20,000 years ago, for instance, glaciers spread across almost 30 percent of Earth's land area, and sea levels dropped low enough to create a land bridge from Asia to North America. But for most of the last 10,000 years, the planet has basked in the milder climate of a so-called interglacial, with average annual global temperatures rising as much as nine degrees Fahrenheit above the average of the last glacial period.

Although scientists are still debating the exact causes of these cycles of dramatic global cooling and warming, one widely accepted theory links them to long-term periodic variations in Earth's orbital geometry, brought on by the gravitational tugs of the Sun, the Moon, and the other planets. For example, the tilt of Earth's rotational axis to the plane of its orbit varies three degrees in 41,000 years. The shape, or eccentricity, of the orbit changes from nearly circular to elliptical in a cycle that takes about

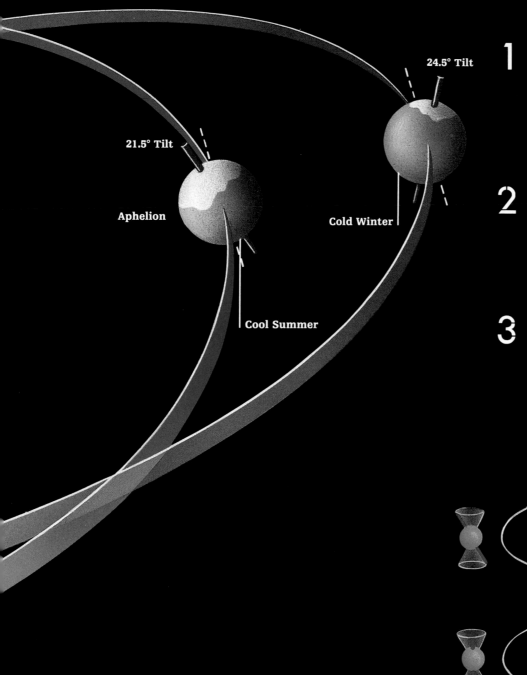

21.5° Tilt

24.5° Tilt

Aphelion

Cold Winter

Cool Summer

1 **Axial tilt.** When the Earth's axial tilt is at its minimum (21.5 degrees), both winter and summer are mild, and ice builds up in middle latitudes. An extreme axial tilt (24.5 degrees) causes cold winters but also makes for hot summers, melting midlatitude snow.

2 **Orbital eccentricity.** Cool summers are also fostered when Earth's orbit is nearly circular *(blue)*. An elliptical orbit *(orange)* heightens the effects both of extreme axial tilt and of precession *(below)*.

3 **Precession.** The slow gyration that causes Earth's axis to describe a cone in space contributes to a gradual shift in the orbital position of the summer solstice. Today, northern summer occurs near aphelion *(below)*. But 11,500 years ago *(bottom)*, summer occurred at perihelion, helping to trigger ice-sheet decay.

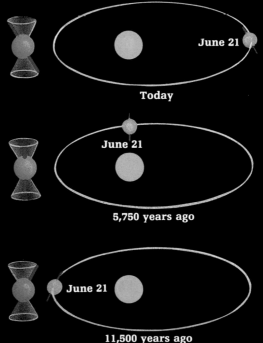

June 21

Today

June 21

5,750 years ago

June 21

11,500 years ago

100,000 years. And in a roughly 23,000-year cycle called precession of the equinoxes *(right)*, the planet is at perihelion, or closest to the Sun, at different seasons of the year.

The interaction of these three cycles affects how much solar radiation the planet receives during any given season at different latitudes. Ice ages seem to occur after many millennia of summers so cool that winter snows remain on the ground, building eventually into glaciers.

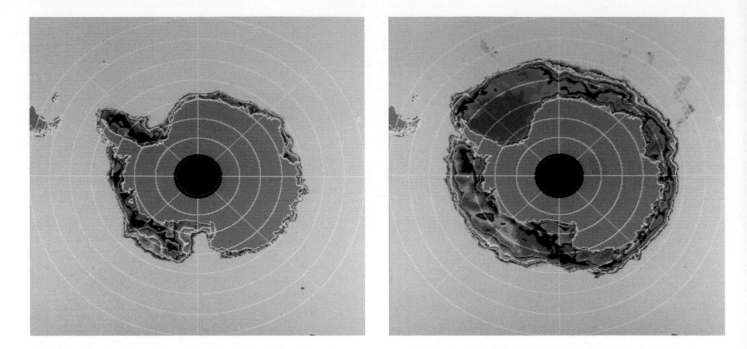

Joined by John Imbrie, an oceanographer at Brown University, Shackleton and Hays published a landmark paper in December 1976. Their examination of the Lamont-Doherty core samples, they wrote, revealed a clear record of ice ages over the past half-million years, occurring at intervals of 100,000, 43,000, 24,000, and 19,500 years—cycles that unmistakably matched those proposed by Milankovitch.

CLIMATES PAST

Although variations in solar energy received by the planet can disturb the machinery of climate, the air-sea engine is largely a sealed unit, capable of running smoothly for billions of years or of backfiring unevenly in times of major environmental change. Today's ocean is a case in point. Below a sun-warmed surface layer, temperatures dive nearly to freezing. An oceanwide barrier of temperature (and density) called the thermocline retards vertical mixing between the surface waters and the denser currents of the abyss, where a typical water molecule spends perhaps ten centuries at a time. Thus the ocean heats slowly and serves as a kind of thermal governor, moderating swings of temperature in the atmosphere from day to day and from millennium to millennium. At the same time, the atmosphere helps to maintain the thermal balance of the ocean. When cold, dry winds blow across the sea, they pick up warmth and moisture, leaving the surface water colder and saltier than usual, which also increases its density. Now heavier than the underlying ocean, this water plunges through the thermocline toward the seafloor, pushing old water toward the surface and replenishing the frigid deeps with nutrients and oxygen.

Scientists believe that almost opposite conditions prevailed when dinosaurs roamed the Earth. The atmosphere was hotter than it is today, as evidenced by the fossils of tropical plants found in Arctic and Antarctic regions, and the oceans were warm all the way to the bottom. Ordinary processes in the sea and air, some climatologists believe, could not have increased average global temperatures enough to sustain a tropical climate

The seasonal waxing and waning of Antarctic sea ice is recorded in four false-color satellite images taken during each of the Southern Hemisphere's seasons, from autumn *(above, left)* through summer *(far right)*. Densely ice-packed waters show up as fuchsia and magenta around the putty-colored continent; lower-density sea ice registers in gold, green, and blue. The periodic freezing and melting of this ice drives the cycle that sends cold, fresh water along the ocean bottom to the equator and draws warm tropical waters south along the surface toward the pole.

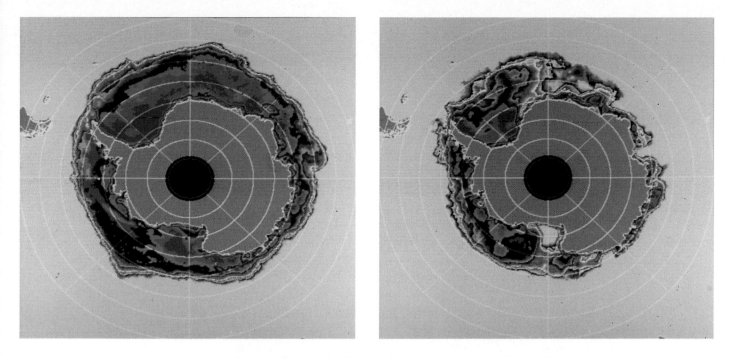

from pole to pole. To account for the necessary heating, scientists hypothesize greater concentrations of carbon dioxide, a heat-trapping compound that one researcher has called the most important gas in the atmosphere. Carbon dioxide admits solar radiation as it comes in but traps the heat energy reradiated by the Earth, creating the temperate climate most of the world enjoys *(pages 104-105)*.

By learning to read ancient climates in fossil remains and sediments, scientists have begun to be able to see beyond the immediate horizon of weather. An early application of this skill came in the 1980s, spurred by a phenomenon that had once seemed so much a Christmas gift from heaven that it was named for the infant Jesus.

YEARS OF ABUNDANCE

Late in 1982, the Lamont-Doherty observatory's research vessel *Conrad* set out from Hawaii for an expedition along the equator, carrying a team of scientists from Woods Hole Oceanographic Institution and North Carolina's Duke University. The Duke team was interested in correlating equatorial currents with nutrients and the abundance of plankton, the small marine organisms at the base of the oceanic food chain. As the *Conrad* began to occupy oceanographic stations—points where the ship stopped and samplers and instruments were lowered to obtain data—the researchers were surprised to find that the surface temperature of the tropical sea was eighty degrees Fahrenheit, five degrees warmer than normal. Stranger still was the nearly complete absence of seabirds, whales, dolphins, fish, and the plankton the scientists had come to study. The equatorial currents, which normally flowed westward, had reversed direction, and the easterly trade winds had died. The scientists aboard the *Conrad* were watching the ocean and atmosphere hatch one of their wilder offspring *(pages 94-95)*.

The phenomenon was nothing new along the coastal desert of Peru. Almost every December, a warm current pushes southward from the equator, warming surface waters. When the current is warmer and stronger than usual and penetrates farther south to Peru, it brings torrential rains to the arid shore. At these times, one traveler wrote in 1926, "the sea is full of wonders, the land even more so. First of all the desert becomes a garden. The soil is soaked by the heavy downpour and within a few weeks the whole country is covered by abundant pasture. The natural increase of flocks is practically doubled and cotton can be grown in places where in other years vegetation seems impossible." Along the Pacific coast of South America, this seasonal gift of moisture was named El Niño—the Child—and the years when it came with special force were known as *años de abundancia*—years of abundance.

THE SOUTHERN OSCILLATION

Even before the 1982 voyage of the *Conrad,* oceanographers had learned that the phenomenon originated in natural events thousands of miles away, driven by changes in winds over the Pacific linked to changes in patterns of atmospheric pressure. Called the southern oscillation, these variations were first detected in the 1920s by a colorful Cambridge-trained mathematician named Sir Gilbert Walker—known to his friends as Boomerang Walker because of his fascination with that aboriginal device. As director general of meteorological observatories in India, Walker became concerned with the causes behind occasional failures in the monsoon circulation, which brings badly needed seasonal rains to the subcontinent. He found that whenever atmospheric pressure was high in the South Pacific, it was low in the Indian Ocean and the monsoon rains were heavy. When pressure in the South Pacific was low, it was high in the Indian Ocean and the monsoon weakened, bringing drought. What he almost certainly did not know was that years of failed monsoon rains in India coincided with Peru's years of abundance.

A generation passed before oceanographers and meteorologists determined that the two events were tied back to back. In 1969, Jakob Bjerknes, who had worked with his father, Vilhelm, on the air-mass theory of weather, concluded that changes in sea-surface temperatures over broad areas of the tropical Pacific produced the observed oscillations of atmospheric pressure. Relatively warm temperatures in the central and eastern Pacific weakened the pressure differences across the ocean, causing the prevailing easterlies to die. Relatively cold temperatures restored strong winds and hence ocean currents. Bjerknes also described the air-sea mechanism that produced the observed effects. A slight diminution of the easterly winds that drive warm surface waters westward can produce a modest warming of the central and eastern tropical Pacific. This in turn can cause a further relaxation of the winds and further warming in a sequence of events that may culminate in strong El Niño conditions. In Bjerknes's view, El Niño and Walker's southern oscillation were both part of "a never-ending succession of alternating trends by air-sea interaction in the equatorial belt."

In 1982, the *Conrad* measured these symptomatic water temperatures as it sailed east along the equator. The warm water was so out of step with climatological norms that a computer back in Washington, D.C., at first

Layers of dust and ice in Peru's Quelccaya icecap reveal some 1,500 years of climatic variation and natural events in the southern Andes. Dust layers contain particles from agricultural fields and exposed lake-bed sediment that may reflect annual water-level changes in nearby Lake Titicaca. By analyzing and dating cores from the icecap, scientists have found a deposit of volcanic ash from a nearby eruption around the year 1600 and have pieced together a history of dry, wet, and ice-age conditions in the region.

rejected the measurements, assuming such data must have come from faulty instruments. But the building El Niño was very real.

Along the coast of Peru, the usual upwelling of cold, food-rich bottom waters ceased, starving the teeming anchovy fisheries that had become a vital element in the regional economy. Seabirds that usually roosted on tropical islands across the Pacific all but vanished. Over the course of 1983, twelve feet of rain brought devastating floods and landslides to South American deserts. Drought struck Indonesia, the Philippines, India, and Australia, where dust storms dumped thousands of tons of topsoil on the streets of Melbourne. In parts of Africa that were already suffering from drought, conditions worsened. Record precipitation in the western United States caused floods in California, the Rocky Mountains, and the Gulf Coast. In normally tranquil Tahiti, which had not experienced a single tropical cyclone in this century, six of the killer storms struck in succession, leaving 25,000 people homeless. By the time El Niño faded, its worldwide toll was estimated to be 1,100 deaths and more than $8.7 billion in property damage.

The experience of 1982-1983 also added a new term to climatology: La Niña, the girl. The term was coined by scientists to identify the other phase of air-sea interactions over the Pacific, when the easterly winds and ocean currents are strong and the surface water relatively cool—the benign conditions that separate the years of El Niño. Sweeter-dispositioned than El Niño, La Niña appears to have a cooling effect on the entire atmosphere and to mark wetter-than-usual monsoon rains in India. But she also has a dark side: La Niña conditions in the Pacific are suspected of triggering dry winters and springs over parts of North America, tightening the grip of regional drought.

FORTUNE TELLERS

The message of El Niño and La Niña is deeply encoded in the air-sea system. According to veteran climate forecaster Jerome Namias of Scripps Institution of Oceanography, "We're dealing with interplay between two very different fluids. Neither medium has a 'normal' state, and abnormality in one causes abnormality in the other. Weather is always abnormal. Perhaps the only thing more complex is human behavior itself."

To see the future of something so complex, researchers have had to merge

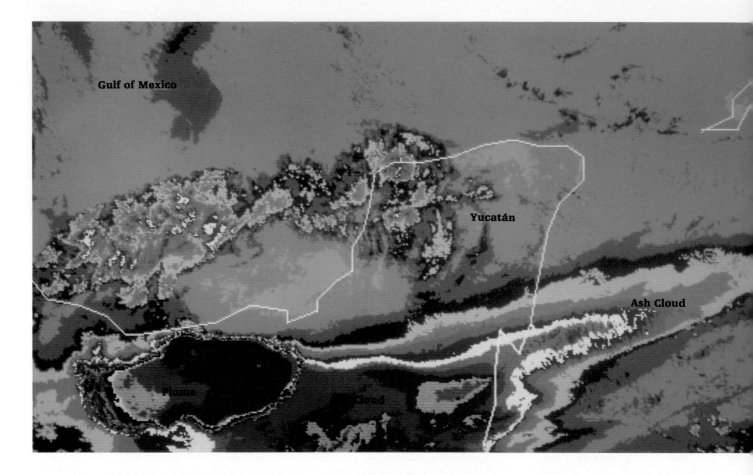

Gulf of Mexico

Yucatán

Ash Cloud

Plume

Cloud

the real world with an imaginary one fashioned from thermodynamic and other equations, more or less along the lines proposed by Lewis Richardson in the 1920s. The difference is that modern atmospheric scientists are able to use electronic rather than human computers, and the crowds in Richardson's forecasting factory have given way to machines processing quantities and equations in what are called numerical models.

In such a model, an imaginary atmosphere is divided into a dozen or more horizontal layers, each latticed into thousands of squares perhaps a hundred miles on a side. The machinery of the model is a myriad of variable quantities, the pulleys and gears of the mathematical factory, all interconnected by a series of equations representing the physics of the atmosphere. As the values of these variables change—for example, as pressure, temperature, humidity, and winds vary—the rates and outcomes of any sequence of events in each of the thousands of grid points composing the model also change. The fundamental test of a model's performance is whether its machinery can duplicate the observed behavior of the real world, if both begin from the same starting point. The more complex a model, however, the more time it requires to run on a computer; even the most powerful supercomputers take hours to simulate a few days' worth of weather.

Climate modelers are less interested in replicating past events than in predicting future ones. If they were modeling a football game, they would want to know enough from the first half of the game to be able to predict events to come in the second. To do this for the atmosphere, meteorologists employ what is called a general circulation model. Requiring billions of

As shown in this false-color satellite image, a colossal ash cloud from the 1982 eruption of El Chichón in Mexico's Yucatán Peninsula carried millions of tons of pulverized rock and sulfuric acid droplets more than ten miles into the atmosphere. At lower levels *(orange and blue)*, the cloud moved to the northeast. But stratospheric winds carried upper layers of the cloud *(red)* southwest and on around the globe, wrapping the planet in a highly reflective shroud. Worldwide air temperatures dropped by as much as one degree Fahrenheit, disrupting trade winds and halting the upwelling of cold, deep ocean waters, which in turn caused the ocean off the coast of California to grow abnormally warm. Major eruptions like El Chichón's offer computer modelers a chance to test their simulations against extraordinary conditions in the real world—for example, to compare modeled and actual global cooling induced by a massive injection of matter into the atmosphere.

calculations, these immensely complicated systems need mountains of rapidly changing information to describe the atmosphere as it is and to predict how it will evolve through time. Feeding their voracious appetites was the reason for such intensive international studies as the one that centered in Africa during 1974.

MERGING AIR AND SEA

Although a general circulation scheme does well enough forecasting large events in the atmosphere over periods of days and perhaps a few weeks, the key to predicting the longer swings of climate is a coupled model—a notoriously complex simulation that merges two very different fluid systems. In addition to the numerical layers of atmosphere, a coupled model includes a shallow bottom layer representing the sea. The first such simulation was developed at the Geophysical Fluid Dynamics Laboratory at Princeton in the 1960s by Sukuro Manabe and Kirk Bryan, mainly as a tool for forecasting weather. When the prototype's predictions were compared with what had actually occurred in the air-sea system, the match was reasonably close. But the model, Manabe recalled nearly three decades later, "could not simulate everything well. In the Atlantic, for example, we always got greater salinity than the real case. This discouraged us. After that, our creativity went backward; it was like the Middle Ages for ten years or so."

Then, as researchers turned to predictions of climate well into the future, the coupled model came into its own. El Niño and La Niña gave scientists an opportunity to demonstrate their ability to predict events on a time scale of months. While neither phenomenon is regarded as the mainspring of the ocean-atmosphere system, both offer researchers an excellent subject for achieving breakthroughs in long-range prediction. Armed with data from the *Conrad* and other research vessels operating in the southeastern Pacific during 1982 and 1983, forecasters were able to predict the occurrence of the next strong El Niño in 1987, using a rudimentary model that is still being improved. They have since begun to merge that tool with larger and more powerful general circulation models, providing a climatological context for the onsets of these alternating phases.

But as forecasters reach toward the twenty-first century and beyond, they run into what seem to be fundamental limits on predictability. The problem is that even the best simulations only approximate reality. Models typically divide atmospheric layers into squares about a hundred miles on a side. Measurements taken at a single point are then averaged over the entire 10,000-square-mile area, which adds error to the simulation. A grid point centered in Los Angeles or London, for example, would ideally include temperature and humidity data not just for the cities but also for the adjacent farmlands and sea; in the real world, these values would be significantly different. Models, however, cannot resolve such fine detail.

To Edward Lorenz, a meteorologist at the Massachusetts Institute of Technology, it is in the unresolved detail that the struggle to predict atmospheric behavior must ultimately be lost. While working on computer models in the 1960s, he discovered what was formally known as the "sensitive dependence on initial conditions." Even slight differences between the model and the real world can produce widely varying results; the less alike they are to begin with, the less likely they are to evolve along similar lines. Because of the uncertainties induced by averaging, a model that seems to reproduce the atmosphere exactly at noon, for example, will be slightly less like the real world one minute later, and will become progressively less true with each passing minute. Over time, the poorly resolved minutiae of actual conditions can evolve into huge errors in a computer model. Lorenz may have expressed his idea most cogently in the title of a 1979 journal paper: "Predictability: Does the Flap of a Butterfly's Wings in Brazil Set Off a Tornado in Texas?" Although the question is unanswerable, sensitive dependence has been known as the butterfly effect ever since.

The work of Lorenz and a host of others has helped atmospheric scientists concentrate on the possible, rather than questing after predictions doomed to failure by the combination of chaos and the butterfly effect. As noted climate modeler Stephen Schneider of the National Center for Atmospheric Research in Boulder, Colorado, has put it, "Climate models do not yield definitive forecasts of what the future will bring. They provide only a dirty crystal ball in which a range of plausible fortunes can be glimpsed."

No matter how dirty the crystal ball or how inscrutable the future, Earth's fortune must be told. If the planet's future is what has been written in its past, the decades and centuries and millennia ahead will bring climatic changes of a magnitude beyond human experience.

OF WIND, WATER, AND WEATHER

From the most tenuous outer reaches of the atmosphere, high above the surface of the planet, to the deepest sea trenches, with their crushing pressures, Earth's double ocean of air and water acts like a giant heat engine, transporting solar warmth from the equator and the tropics toward the ice fields of the poles, and drawing cool air and water toward the equator. An intricate system of currents, spawned by these temperature differences and modified by Earth's rotation, carries out this ceaseless heat exchange.

By moving masses of air and water over Earth's surface, the currents give rise to the patterns of prevailing winds and ocean currents that have in turn determined human patterns of agriculture, fishing, and transportation for thousands of years. At the same time, inexplicable hiccups in the system can produce destructive anomalies—devastating floods in normally dry regions or seasons, drought that lasts for months on end, the disappearance of ordinarily plentiful fish from coastal waters.

As illustrated on the following pages, the interaction between the Sun's radiant energy and Earth's fluid envelope of air and water creates day-to-day weather and, in the long run, climate. So complex is the process, however, that scientists have only partly unraveled the secrets that give the Third Planet its chameleon face of blue sea and white clouds.

electromagnetic spectrum, from high-energy, short-wavelength gamma rays to the long waves of radio, most of what gets through to Earth is infrared and visible light *(right)*. The rest is reflected, absorbed, or scattered at various levels of the atmosphere.

At about 100 miles up, incoming solar radiation encounters the thermosphere, a rarefied layer of atomic oxygen and nitrogen that extends down to about 50 miles above sea level. The atoms absorb the most energetic radiation—gamma rays and x-rays—which raises temperatures in the near vacuum of the thermosphere as high as 3,600 degrees Fahrenheit. The process strips electrons from the oxygen and nitrogen atoms, leaving them ionized, or electrically charged; hence this region, which blocks many radio wavelengths, is also known as the ionosphere.

At an altitude of about fifty miles, the remaining sunlight reaches the mesosphere (*mesos* is Greek for "middle"), a domain of molecular and atomic nitrogen and oxygen too thin to absorb much radiation; it thus stays chilled below minus 140 degrees. Most of the radiation passes through this layer to the stratosphere twenty miles below. In this denser region, short-wave ultraviolet light tears oxygen molecules into individual atoms, some of which recombine to form the three-atom form of oxygen called ozone. The ozone soaks up most of the incoming ultraviolet radiation, which would be lethal to many life forms if greater doses of it reached the surface. The absorbed radiation raises temperatures in the stratosphere to between minus 50 and minus 10 degrees.

At ten miles up, sunlight enters the troposphere, province of most weather. Water vapor, carbon dioxide, trace gases, and particles of dust and pollutants absorb infrared radiation and are heated by it. Tropospheric air molecules and microscopic dust particles scatter blue wavelengths to color the sky.

Earth's fluid mantle extends nearly 300 miles, from the thermosphere *(top)* to the ocean deeps. Upper levels of the atmosphere block most high-energy components of sunlight, but some infrared and almost all visible radiation reaches the surface, where it is stored as heat. In the ocean, a 600-foot-deep transition zone insulates warm surface water from the chill abyss.

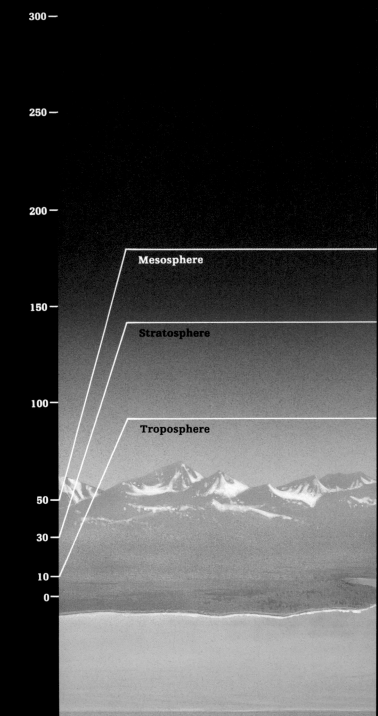

Mesosphere

Stratosphere

Troposphere

Gamma   X-Rays   Ultraviolet   Visible Light   Infrared   Radio

A GLOBAL HEAT PUMP

Vertical and horizontal transfers of heat, most of which take place in the few miles of air and water closest to sea level, modulate Earth's climate. In that narrow region, solar energy stored in the uppermost layer of the sea warms the air from below, just as a gas ring warms a pot. Some of this heating occurs directly, by contact between the ocean and the overlying air. More than twice as much heat is moved indirectly, through evaporation of ocean water.

Evaporation occurs when molecules close to the surface acquire enough energy to escape the liquid altogether. When water turns to water vapor, the surrounding ocean loses heat energy and the evaporating water gains the heat in a stored, or latent, form.

Carried aloft aboard rising molecules of water vapor, latent heat does not warm the air around it. However, as the humid air ascends, it encounters lower atmospheric pressures, causing it to expand and cool, which reduces its ability to hold water vapor. At a temperature called the dew point—a function of atmospheric pressure and the amount of water vapor present—the air becomes saturated and the invisible water vapor begins to condense on particles of dust, sea salt, pollen, and pollutants drifting in the atmosphere, forming tiny droplets that accumulate into clouds. Because condensation reverses evaporation, the condensing water releases its latent heat into the cloud.

This causes the currents in the cloud to rise farther, releasing still more heat as water droplets are lifted above the freezing level in the atmosphere and turned into ice crystals. As more water vapor is swept up into the evolving cloud system and condenses or freezes, additional heat is released and the cloud system continues to grow. Eventually, however, the cloud droplets either coalesce into large drops or freeze and grow into snowflakes. In either case, they become heavy enough to fall, returning the moisture borrowed from the sea.

Heated by the Sun, some ocean water molecules *(gray)* acquire enough energy to fly into the air. This energy is called stored, or latent, heat of vaporization.

Water vapor rises and cools, condensing into droplets *(blue)* to form a cloud. Latent heat *(pink)* released in the process warms the surrounding air.

Some of the warmed air rises still higher, extending the cloud upward into a vertical tower, and carrying water droplets with it.

High in the cloud, droplets freeze to ice crystals, releasing more heat *(pink)*. As the ice falls to warmer altitudes, it melts into rain.

THE ORIGINS OF PREVAILING WINDS

The power that drives the air-sea heat engine is the unequal heating of Earth's surface resulting from the curve of the globe itself and from the planet's axial tilt toward or away from the Sun, determined by the position of Earth in its orbit. In the tropics, sunlight strikes directly, through less atmosphere, almost all year round; but at middle latitudes, the Sun's rays are more oblique, especially in winter. The polar regions feel only a glancing touch of solar power during part of the year.

Because the tropical ocean absorbs considerably more heat than waters at higher latitudes, convection currents arise in the atmosphere *(opposite)*. Buoyant, ocean-heated air near the equator soars up from the surface and drifts toward the cold polar regions. Cooler, denser air in the subtropics and the polar regions sinks, moving toward the equator to replace the warm rising air. Because of Earth's rotation, moving air is deflected to the right in the Northern Hemisphere and to the left in the Southern Hemisphere, a phenomenon known as the Coriolis effect *(diagram, above, right)* for its French discoverer. In the northern subtropics, the warm equatorial air gradually cools and sinks, dividing into two currents, both of which are also deflected. Air currents that become the comparatively steady easterly trade winds move south to replace the rising air at the equator; more variable westerlies, with embedded swirls and eddies, flow north. A mirror image of this process occurs south of the equator.

Continents and islands, which act as barriers to the flow of ocean currents, also affect airflow at low altitudes. Landmasses respond more rapidly than the oceans to temperature changes, cooling down faster at night and in winter and warming up faster during the day and in summer. Such differences may deflect or even reverse air currents as they move from sea to land. The combined effect twists air currents into a series of bent spokes radiating from the central oceans in subtropical latitudes *(lower right)*.

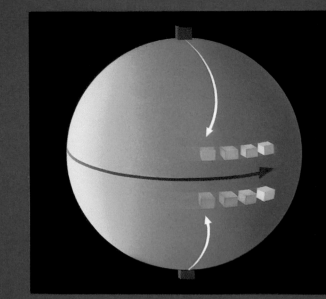

As illustrated above, the Coriolis effect is a function of Earth's rotation. A hypothetical object thrown from the North Pole toward an observer in the northern tropics would appear to the observer to have curved to the west rather than following a straight north-south line, because the planet's spin carried the observer east while the object was en route. In the Southern Hemisphere, an observer would perceive a mirror-image effect.

Large-scale winds swirl away from the center of each ocean in this map of typical air currents during the month of July. Because the conditions that produce wind patterns vary with the annual cycle of solar heating, every month has a different characteristic wind pattern.

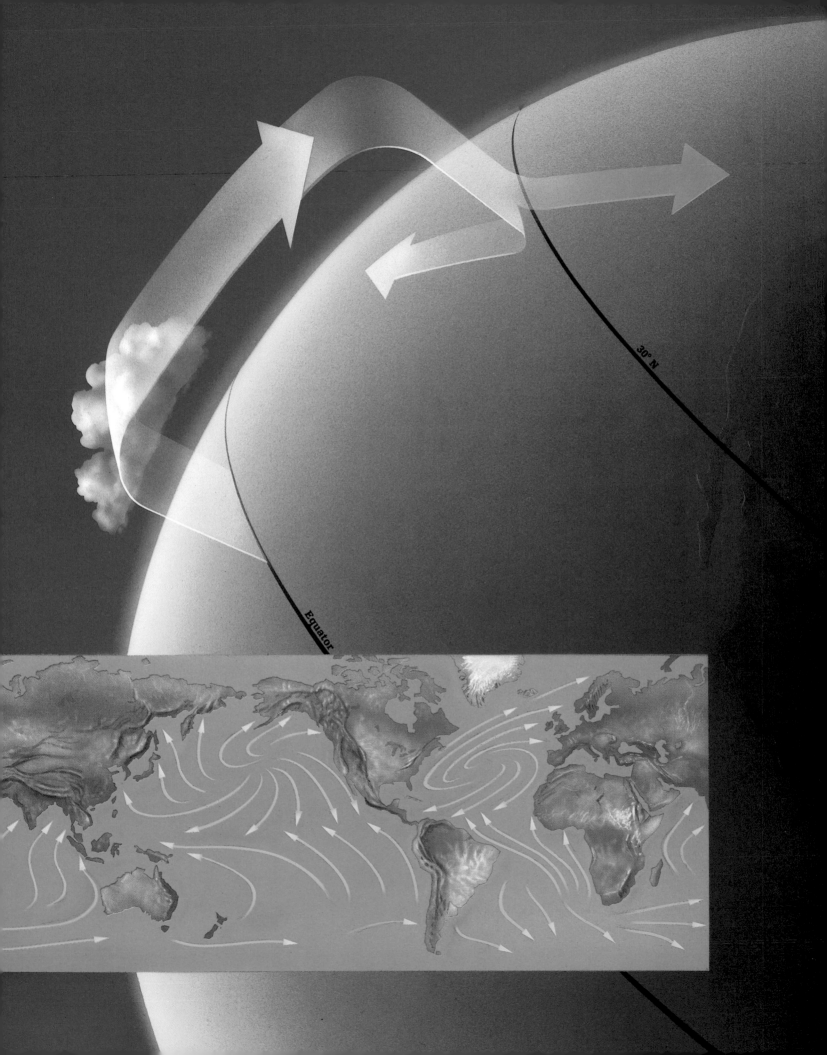

30° N

Equator

Rivers in the Sea

Nearly half the energy transported from the equator toward the poles is carried on vast, warm currents in the surface layers of the ocean. In the North Atlantic alone the Gulf Stream transports more energy northward every hour than would be produced by burning five billion tons of coal.

These oceanic rivers are shaped by a combination of prevailing winds and the planet's rotation. Winds push the surface currents, which are deflected somewhat by Earth's rotation and by friction between air and ocean. The deflection causes seawater to accumulate at central points in each ocean, forming mounds several feet high. Pulled by gravity, the piled-up water continually spreads toward the sides of the ocean basin, spun by the Coriolis effect into circular currents called gyres.

Planetary rotation also pushes the center of each gyre to the west. Squeezed by the distortion, the currents along the western sides of gyres are narrow and fast. In the Northern Hemisphere, the waters of both the Gulf Stream in the Atlantic and the Kuroshio (Japanese for "black tide," after its ultramarine color) in the Pacific have been clocked at over sixty miles per day. Meanwhile, the eastern sides of gyres broaden into wide, slow-moving streams like the California and Canaries currents, which return cooled water to its tropical heat source.

Earth's rotation and prevailing westerly winds in high southern latitudes also give rise to the Antarctic Circumpolar Current, the only ocean stream that is not hemmed by continental masses and can thus circumnavigate the globe. Uniformly chill polar waters permit the Antarctic current to extend two miles down, transporting a greater tonnage of water than any other surface current.

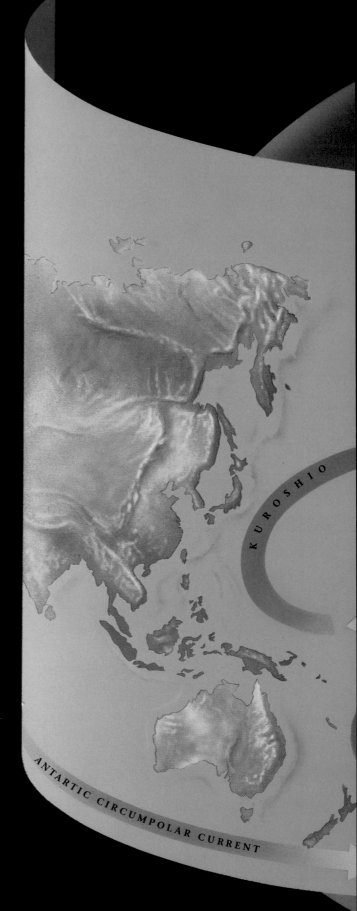

KUROSHIO

ANTARTIC CIRCUMPOLAR CURRENT

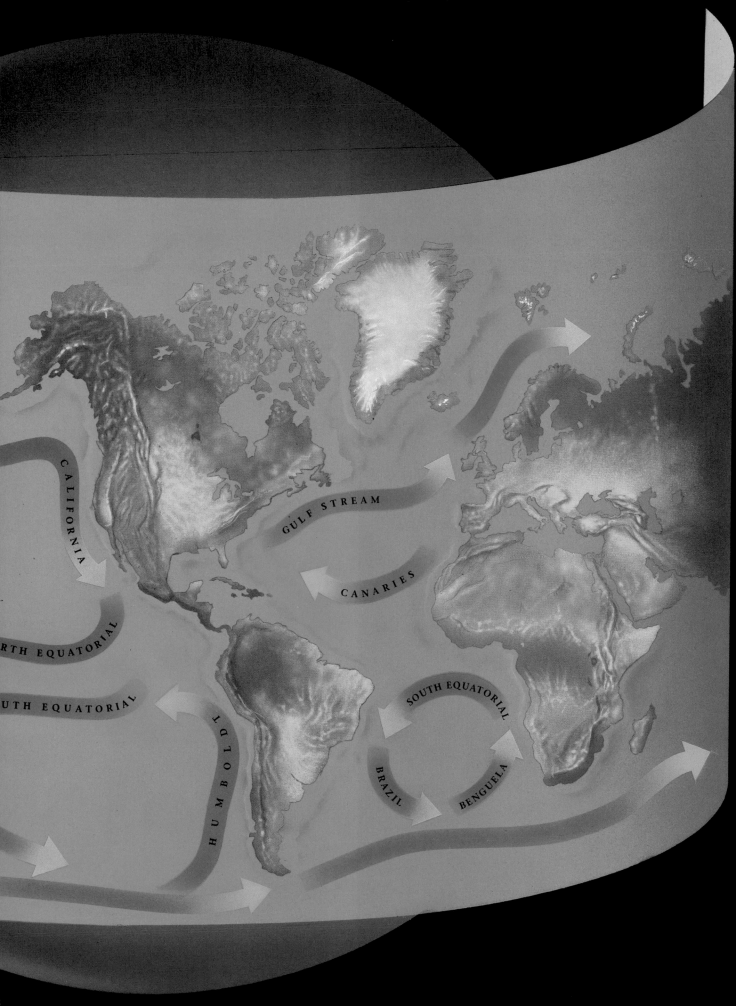

The Enigma of El Niño

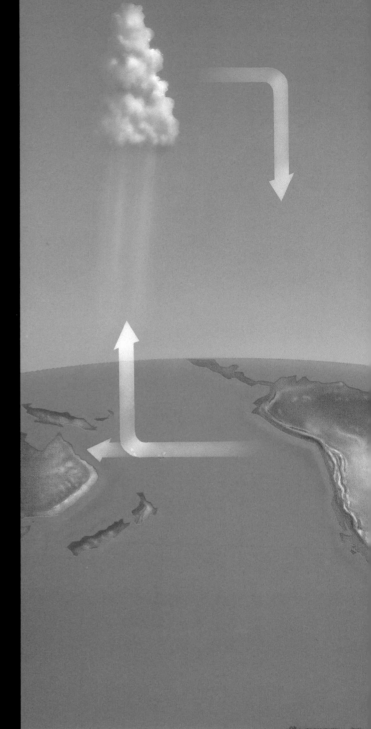

Once believed to be a regional occurrence, the phenomenon known as El Niño is now recognized as a global fluctuation in normal weather patterns. Named for the Christ child, because it occurs around Christmas every few years, El Niño involves a series of complex interactions between the atmosphere and the ocean, making it difficult to predict. Warm surface water spreading eastward from the central Pacific brings lower than normal atmospheric pressure to the area off Ecuador and Peru. This causes the trade winds to falter, which in turn allows more warm water, piled up in the western Pacific by the trade winds, to flow eastward to South America, diminishing the supply of cold, nutrient-rich water from the deeps. The reduced pressure and warmer coastal waters bring heavy rains to the normally dry west coast of South America. Conversely, unusually high pressure over the Indian Ocean brings drought conditions during the normally rainy seasons in India, Southeast Asia, and Australia.

The irregular alternation between these El Niño and the tropical opposite conditions, now called La Niña (the girl), is known as the southern oscillation. In years of La Niña, high pressure over the eastern Pacific keeps the trade winds strong. Peru's deserts remain dry, upwelling cold waters supply food for the coastal fisheries, and on the other side of the Pacific, the monsoon revives southern Asia with vital rains.

The two pairs of illustrations at right show how El Niño disrupts the usual patterns of atmosphere *(top)* and ocean *(bottom)*. Normally *(near right),* high pressure in the eastern Pacific keeps the westward trade winds strong, pushing warm surface water west and allowing food-rich cool water to well up near Peru; low pressure in the western Pacific brings rains to Asia. Pressure weakens and sometimes reverses during El Niño *(far right).* Trade winds are intermittent; rain falls largely in the Pacific instead of Asia, and cold waters no longer rise.

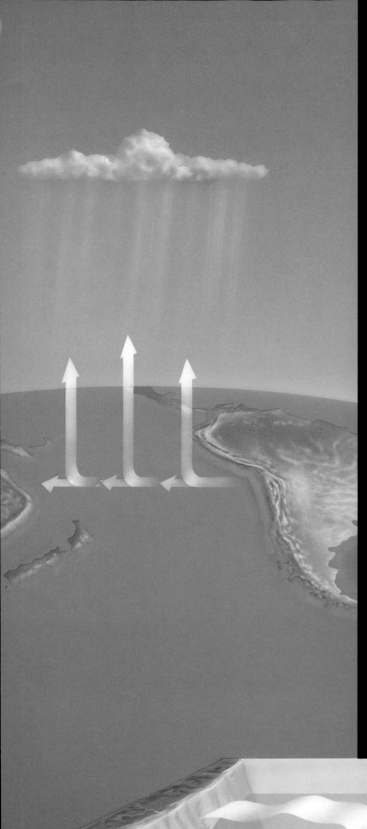

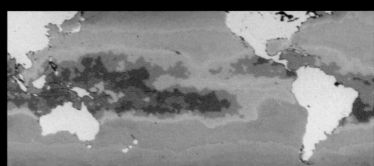

In normal years, a tongue of cool water *(green)* wells up along the Peruvian coast and extends westward into the warm equatorial Pacific.

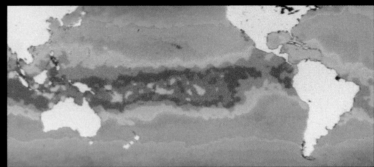

During the 1982-1983 El Niño, warm western waters *(yellow, red, magenta)* sloshed east, and the cool equatorial upwelling vanished.

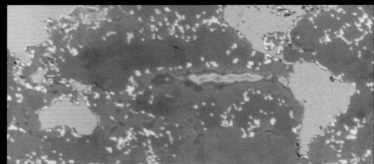

Temperature differences between the two readings, with orange indicating warmer than normal, show the zones most affected by El Niño.

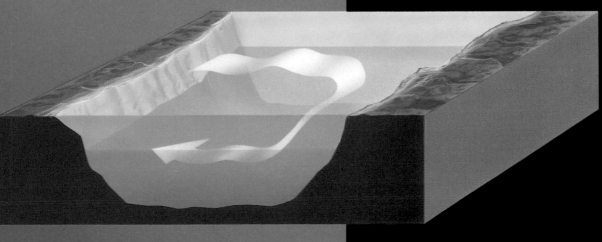

4/PORTENTS OF CHANGE

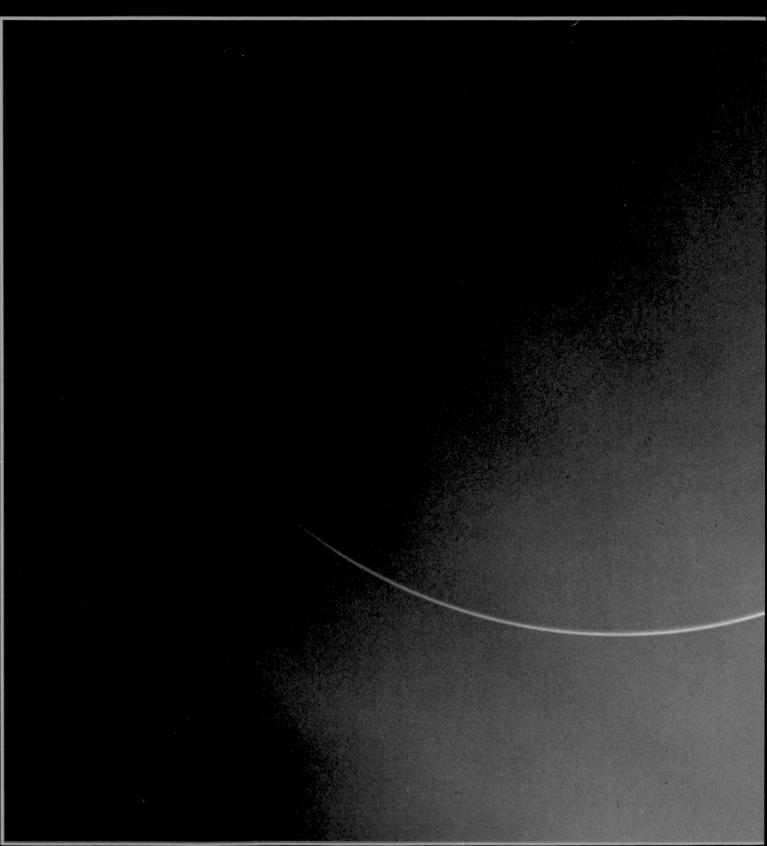

on their return from the Moon in November 1969. The global perspective offered by spaceflight has fueled speculation about how the planet will evolve.

he island of Jamaica seemed a paradise to the exhausted men aboard the three Spanish caravels that approached its shore in July of 1494. Under the leadership of Admiral Christopher Columbus, they were exploring the Caribbean, but instead of finding gold and precious stones, they had been engaged in a protracted struggle for survival as their stores dwindled and storms battered their ships. Only a week before, the admiral had written in his log, "Not a day passes that we do not look death in the face." But now their luck appeared to have turned. The crew exulted at the sight of the Jamaican coastline, where dark, misty forests ran down to the water and cloud-wrapped peaks rose high into the warm tropical winds. To Columbus, these elements seemed somehow to be linked. His son later reported in his account of his father's voyages: "The sky, air, and climate were just the same as in other places; every afternoon there was a rain squall that lasted for about an hour. The Admiral writes that he attributes this to the great forests of that land; he knew from experience that formerly this also occurred in the Canary, Madeira, and Azore Islands, but since the removal of forests that once covered those islands, they do not have so much mist and rain as before."

Columbus, a man of little science but vast knowledge of the sea, had sensed that Earth's life forms and climate are intertwined, producing subtle yet important changes in one another. Most modern scientists concur. But exactly how living things connect to the physical world is still unknown.

In recent years, much detective work has been directed at understanding the sort of relationship that Columbus so astutely observed. During the autumn of 1985, for example, a major field experiment began in the humid basin of the Amazon River and its dome of hot, hazy atmosphere. For six weeks in the tropical dry season, teams from the National Aeronautics and Space Administration and Brazil's National Institute for Space Research probed the steamy respiration of the rain forest and the great river that crosses and feeds it. A battery of instruments mounted on meteorological towers and carried by planes, river-going ships, and balloons analyzed the chemical breath rising from the forest and the Amazon. Part of NASA's Global Tropospheric Experiment, the Amazon project was the first of a series of projects exploring the links between biology and climate. Subsequent missions brought scientists and instruments back to the Amazon for six weeks of the 1987 wet season, when powerful thunderstorms mixed the chemicals released by the rain forest high into the atmosphere.

In 1988, the NASA researchers moved to the tundra of northern Alaska, and they will extend their work further across the top of North America in 1990. They also plan studies of what happens to the Amazon atmosphere as it travels over the equatorial Atlantic, and expeditions to the Pacific, where, according to project director Robert J. McNeal, they will find some of the cleanest air on the planet. The main thrust of these expeditions is to lay down baselines, or norms—the chemical behavior of an undisturbed environment. But the studies are also predictive: By describing a geophysical present for these remote locations, McNeal and his colleagues hope to establish a backdrop on which scientists will be able to read the faint signals of global change. Even subtle trends should be seen dramatically and quickly in these fragile, relatively untouched areas McNeal calls "the edge of the delicate climate machine." A fluctuation in the depth of permanently frozen ground at high latitudes, for instance, is a clearer signal of temperature trends than the abundant, but often conflicting, data available for temperate latitudes.

The stakes are much higher than the sort of local effects observed by Columbus. Today's question is whether and to what degree human activities will alter the climate of entire regions and possibly the whole globe. Answers to such queries are being fashioned into a rough map for humankind's expedition into the future. The craft is the good ship Earth itself, with some five billion people aboard and uncounted multitudes of other species, plant and animal. Many experts see alarming climatic possibilities ahead—profound and rapid changes in the composition and behavior of the planet's atmosphere, and in the nature and number of living things that cling to its surface. Other than the informed intuition of scientists, however, the only reliable charts for this journey are those of climates past.

MACHINES OF CHANGE

At present, no one fully understands why Earth's climate varies. External forces—fluctuating amounts of the Sun's energy, for example, or volcanic carbon dioxide added to Earth's atmosphere—plainly have an effect. But change also appears to be an inherent attribute of climate, not just a reaction to factors imposed from outside the air-sea system. For example, scientists have come to recognize that huge portions of the Pacific and Indian oceans, and the atmosphere above them, alternate between the two different states that have come to be called El Niño and, more recently, La Niña *(pages 94-95)*. Both the timing and the magnitude of the phenomenon vary considerably. The 1982-1983 El Niño was far more intensely developed than any other in recent decades. But in the flood patterns of Peruvian canyons, archaeologists have found evidence of monster El Niños perhaps ten times stronger; fifteen of these are known to have left their fossil signature in the past 10,000 years. The distinct but irregular pattern of their occurrence indicates that the El Niño oscillations are not driven by external forcing; they are an integral part of the climatic process.

On a grander scale of change, Earth's geologic record shows periods of glaciation called ice ages. These have been linked to eccentricities in Earth's orbit, its tilt, and the precession of its axis: the Milankovitch mechanism *(pages 76-77)*. But some climatologists now argue that the resulting shifts in solar heating are not the sole cause. They may trigger the glacial cycle, but once the change is set in motion, it continues to evolve because of internal processes. For example, a warming trend initiated by orbital fluctuations might be reversed as continental ice sheets melted, abruptly diluting the ocean with huge quantities of fresh, cold surface water.

Other long-term effects on climate may come from slow shifts in the Earth's crustal plates. As the continents drift, they force rearrangements in the patterns of ocean currents, thus changing the way the oceans distribute heat around the globe. Collisions between continents create new mountain ranges, altering circulation patterns in the lower atmosphere. Similarly, because landmasses are the foundation on which polar ice sheets spread, their positioning plays a major role in regulating the flow of energy received from the Sun. Fifty million years ago, for example, before Antarctica moved to its present location, the South Pole was not covered by a large icecap. Without this powerful attractor, equatorial heat did not flow to the south, and the strong belts of prevailing winds and ocean currents were weak and intermittent compared with today's.

A CHILLY WORLD

Because climate varies so slowly, it spins an illusion of great stability. But within broad limits, change, sometimes rapid change, has occurred again and again, punctuating more static epochs. Climatologists cannot cite a climatic norm for the simple reason that there is none—or rather, there have been many.

Today, the average surface temperature of Earth is about fifty-seven degrees Fahrenheit. When the dinosaurs lived, the equivalent temperature was perhaps seventy-two degrees. There were no glaciers then, and the polar regions were largely ice free. Between such balmy periods, however, the planet has tended to be a good deal colder, resulting in the spread of mile-thick ice sheets as far south as the latitudes of present-day Boston and London. In the past billion years, there have been at least four separate epochs when extensive portions of the globe were covered by ice, separated by extended interglacial periods when the ice sheets retreated.

Past evidence would suggest that Earth is now in a period of global cooling that began about 50 million years ago, a slow descent into another period of glaciation punctuated by intervals of warmth. It has been several million years since Earth's temperate zones were last free of ice, although partial thaws lasting about 10,000 years occurred every 100,000 years or so during the past million years. In the latest of these, a relatively warm period beginning just 10,000 years ago, human civilization arose.

The climate has fashioned many brief variations upon the larger theme of advancing and retreating ice sheets, becoming fitfully warmer, colder, wetter, or drier. About 5,000 years ago, for example, the once-lush plains of North Africa and Arabia dried up: Now the shifting sands of the great deserts cover former streambeds. During the first-century campaigns of Caesar's legions, Britain and Germany had mild climatic conditions. But a little over a thousand years later, from about 1500 AD until the mid-1800s, the Northern Hemisphere shivered through what is called the Little Ice Age, when crops froze at high latitudes and famine was common. Since then, global climate has been generally benign, with a slight cooling in the Northern Hemisphere between 1940 and 1975 and an apparent warming trend in the 1980s.

As it happens, however, the prospect of thick ice sheets planing the Northern Hemisphere is not what worries climatologists today. Between the present point in time and the next era of ice, scientists expect a global warming of a magnitude that many of Earth's life forms have not experienced before. Some researchers see a bleaker future yet: a planet smothering in its own atmosphere, trapped in an escalating spiral of heat.

A BALANCED BUDGET

Whatever its degree, the predicted warming begins with the radiant energy received from the Sun. Astronomers comparing the Sun to other, similar stars at different stages in their life cycle estimate that it is now perhaps 30 percent brighter than when the Solar System was formed 4.6 billion years ago. Over the next billion and a half years, the Sun is expected to become about 15 percent brighter than it is today. As its output of radiant energy increases, Earth presumably should heat up. But the climatic record shows the process to be more complicated.

What counts, ultimately, is not how much of the Sun's radiant energy reaches Earth, but how much the planet keeps. Normally, as much energy is reradiated to space as is received from the Sun *(pages 102-103)*. About 30 percent of the Sun's radiation is immediately reflected off ice or clouds and lost to space. The rest is absorbed—by the atmosphere, oceans, plants, and soil. As the molecules of these materials heat up, they reradiate the absorbed energy, which eventually finds its way back into space, unless something intervenes. That something is primarily molecules of heat-absorbing gases like carbon dioxide and methane, which trap the heat radiation rising from the planet, preventing its escape. A slight shift in the atmospheric concentration of such gases, then, can alter the climate.

While the general composition of air—78 percent nitrogen, 21 percent oxygen, 1 percent argon—has remained essentially unchanged for more than two billion years, the atmosphere's burden of trace materials alters constantly *(pages 104-105)*. Carbon dioxide molecules enter the air when volcanoes erupt and when animals respire or organic matter decays. The molecules are removed from the air by the photosynthetic activity of plants and when they fall to the sea and soil in raindrops. Geochemical or biogeochemical cycles also continuously alter concentrations of these gases.

Earth receives some 130 trillion horsepower of energy from the Sun every second. And when the planet is done with it, an identical amount is sent back into space. The entire planet—its air, oceans, winds, even its carpet of living things—contributes to this grand exchange. About 30 percent of the incoming radiation *(yellow arrows, below)* is bounced directly back to space. The remainder is absorbed by the planet in some fashion: 25 percent soaks into the atmosphere and never hits the ground, 22 percent penetrates to

the solar radiation absorbed by the planet is reradiated as infrared *(red arrows).* Four percent (of the original 70 percent) returns unimpeded to the void; the rest follows after some back-and-forth interactions with the atmosphere.

In this way, Earth balances its energy budget, maintaining a global average air temperature at the surface of some fifty-seven degrees Fahrenheit. But the balance is delicate and could be tipped by a small

Reflected by Surface (5%)

Reflected by Atmosphere (25%)

Absorbed Directly by Surface (22%)

Absorbed by Atmosphere (25%)

Absorbed by Surface through Atmospheric Downscattering (23%)

change in either of two key factors: albedo and the greenhouse effect. Albedo is a measure of the energy reflected by the planet's atmosphere and surface. Snow, for example, reflects as much as 80 percent of the light that reaches it; oceans reflect an average of 10 to 15 percent of incoming light. If Earth underwent a measurable change—if, say, deserts were to spread over a greater surface area than they now cover—the planet would reflect more light, thereby lowering its average global temperature.

The greenhouse effect warms the globe by preventing the escape of infrared radiation. Certain gases that are present in the atmosphere, such as water vapor, carbon dioxide, and methane, are fairly transparent to incoming short-wavelength radiation but relatively opaque to the outgoing longer-wavelength infrared. As a result, most infrared from the surface is absorbed by atmospheric gases and particles and reradiated in all directions, heating the ground even more. After a series of such surface-to-air exchanges, enough energy leaks into space to balance the absorbed solar input, but not before blanketing Earth with a layer of warm air.

A change in the amount or composition of atmospheric gases—an increase in the percentage of carbon dioxide, for example—would trap more heat around the planet, causing global surface temperatures to rise. Temperatures would stabilize eventually, but at a higher level.

Emitted by Surface (4%)

Emitted by Atmosphere (66%)

CHEMICAL RECYCLING ON A GLOBAL SCALE

Albedo and atmospheric composition, which influence global climate change, are in turn governed by processes that endlessly cycle chemical nutrients through the environment of earth and sea and back into the atmosphere. The hydrological cycle, for example, moves water through Earth's system. Another process does the same for nitrogen: Plants take up nitrogen from the soil; when they decay, bacteria unlock the element and release nitrous oxide, among other gases, to the air, where it helps to hold heat near Earth's surface. Sulfur is another element that cycles into and out of the atmosphere. Sulfur particles in the atmosphere can form a haze that blocks sunlight or makes clouds brighter, thus changing the planet's albedo. As sulfuric acid, it also affects the growth of Earth's plants and the ecological balance of its bodies of water.

One of the cycles most critical to terrestrial climate involves carbon, the building block of all organic molecules. As shown below, it travels through nature in an efficient round of depletion and renewal. Plants draw in carbon dioxide during sunlight hours and release a portion of it at night (1). When the plants die, carbon enters the earth and is recycled to the atmosphere through bacterial action. If the decaying tissue is

trapped in an oxygen-free setting, such as a swamp, some of it may become compacted over the eons to form oil, gas, and coal deposits (2)—which humans might subsequently tap and burn, returning carbon to the air (3). As part of the atmospheric gases carbon dioxide and methane, it acts as a barrier to infrared radiation, trapping heat and warming the planet. Carbon also seeps into the ground in the form of carbonic acid in rainwater (4), reacting with more carbon and minerals in Earth's crust to create calcium carbonate and silicic acid. These travel with ground water to the sea, where marine animals use the carbon to build their shells. When the organisms die, the shells fall to the ocean floor and are compacted into limestone (5). Eventually the limestone will be subducted, pressed down into the layer known as the mantle as the oceanic plate in which the limestone is deposited encounters a continental plate (6). Carbon finds its way to the air again as carbon dioxide through volcanoes (7) or into the ocean through midocean ridges (8).

Meanwhile, carbon in the air dissolves easily in cool seawater, reentering the atmosphere when warm tropical waters reach the saturation point and release it (9). Plankton and other tiny marine organisms also contribute to the exchange of carbon between sea and air, absorbing carbon dioxide by day and returning part of it by night (10). If even one variable in that cycle or in the others changes, the whole atmospheric equation may shift, bringing a new climatic regime upon the Earth.

Although the heat-trapping action of such compounds is not the same process seen in greenhouses (whose interiors are warm because Sun-warmed air trapped inside cannot circulate enough to cool), they are called greenhouse gases. Without their warming effect, Earth would have frozen early in its history and likely remained a dead planet.

One other naturally created gas plays an important role in Earth's climate: water vapor, the volatile wild card of the atmosphere. Like carbon dioxide and methane, water vapor is a greenhouse gas. It absorbs infrared radiation in a different part of the spectrum than carbon dioxide and is less effective per molecule in trapping heat, but this inefficiency is compensated by the huge amounts of it—some 130 billion trillion gallons—in the atmosphere. The warmer the Earth becomes, the more moisture enters the atmosphere and the more heat it traps. But nothing about the behavior of water is simple. If it warms by trapping more heat, it also cools by creating more clouds, whose tops, like mirrors of ice, reflect sunlight. Scientists believe that, at the moment, the net effect of water vapor is a slight cooling: Clouds reflect more energy than atmospheric water vapor traps, but just barely. Whether

An intense bloom of plankton swirls clockwise in coastal waters off New Zealand in 1985. The microscopic plants may help moderate climatic extremes. According to one theory, when sea-surface temperatures rise, plankton may release increased quantities of sulfur compounds, around which water droplets form, encouraging cloud formation. Increased cloud cover reflects more sunlight, cooling the surface back to normal.

that will continue to be true if temperatures and cloudiness increase, no one can say. From a scientific point of view, water in the atmosphere remains a climatic enigma.

Part of the mystery rests in how clouds form. The process begins when water vapor condenses into droplets. Without something to condense on, however, vapor tends to remain a gas. Scientists have long assumed that windblown dust particles in the atmosphere served as tiny platforms called condensation nuclei, where water vapor would readily change into its liquid or solid state in a process called nucleation. But the origins of condensation nuclei, and the detailed physics of clouds, are still not fully known.

One source is clear enough. Satellite observations of clouds over shipping lanes have shown that particles from ships' smokestacks can act as condensation nuclei. Because smokestack particles are smaller and more numerous than natural ones, the clouds they create are especially bright and reflective.

A more controversial source for some maritime clouds was put forward in 1987: marine organisms. According to British inventor-theorist James Lovelock and a University of Washington team led by Robert Charlson, a specialist in atmospheric particle chemistry, dimethylsulfide produced by plankton escapes into the atmosphere, where it reacts with oxygen to form particles on which water can liquefy. Because the droplets collect into clouds, which reduce the amount of sunlight reaching the surface, the plankton, in effect, cool the water in which they live. The theory has not won universal acceptance, but it is taken by some researchers as proof that living organisms and climate affect each other.

It also tends to support the Gaia hypothesis, proposed in the 1960s by Lovelock and his American colleague Lynn Margulis. Named after the Greek goddess of Earth, the Gaia hypothesis argues that Earth is a self-regulating system; that is, living organisms have altered the planet in ways that keep it suitable for life. The oxygen in the atmosphere, for example, is cited as evidence of Gaia: It is so quick to react chemically with other substances that it would disappear into a multitude of oxygen-containing compounds if it were not continuously replaced by plants and algae. Such biological effects pervade the physical world. Bacteria in the soil hasten the decomposition of organic matter and even rocks, recycling their minerals into Earth's chemistries. Plankton, which may modify regional climate, also figure prominently in geochemical cycles. They take up carbon from the oceans to form their skeletons and bury it in the seafloor sediments when they die, locking excess carbon into oceanic plates that will eventually return to the Earth's mantle. Thus it is life itself, according to Lovelock and Margulis, that has kept the planet vital.

A contrasting view holds that Earth's climate is purely an accident of birth, resulting from celestial position, mass, and the inanimate forces of physics and chemistry. With or without life, NASA researchers James Kasting, Owen Toon, and James Pollack have argued, the process of weathering would gradually remove carbon dioxide from the atmosphere and bury it in the ocean sediments. Over the course of geologic time, the movement of crustal plates would then drag those sediments into the planet's hot interior, eventually liberating carbon dioxide in volcanic eruptions. In the view of these and other scientists, this type of geophysical cycle—and the luck of being far enough from the Sun to avoid an accelerating greenhouse effect—explains Earth's conditions without the need to invoke hypotheses like Gaia.

Perhaps, as proposed by Stephen Schneider at the National Center for Atmospheric Research, the truth lies somewhere in between. Schneider and others believe there is a climatological middle ground where both life and geophysical processes are potent but neither dominates. Instead, the forces have coevolved. As Schneider has put it, "The physical, chemical, and biological subcomponents of the earth all interact and, whether by accident or design, mutually alter their collective destiny." Thus the purely geophysical mechanisms may, over millennia, cause the composition of the atmosphere to vary, but the cycle is modified by living organisms in unpredictable ways. For example, the oxygen-rich atmosphere produced by living things means that fires can occur, providing a new and sometimes dramatic means of recycling carbon back into the atmosphere. And in the past few hundred years—a mere eyeblink of geologic time—one form of life, humans, has learned to mine carbon stored in the planet's crust and burn it, creating huge additional quantities of carbon dioxide and another level of uncertainty.

THE HUMAN FACTOR

Human behavior has become the great imponderable in predicting climatic change. Evidence of its impact abounds. Thousands of fires set to clear land for farming and grazing pour smoke across the green canopies of rain forests. Over industrialized regions, a fine haze of smog absorbs incoming sunlight and outgoing heat. In some places, the natural contours of the land act as huge basins that concentrate the smog, causing whole cities to vanish in a murky, choking veil. In others, the smog blends with ordinary clouds, thickening the cover over wide areas marked by forests that are already stunted or killed.

The geologic record shows that, over the long span of the planet's life, the intensity of the greenhouse effect has varied greatly, governed mostly by the amount of carbon dioxide in the atmosphere. If, as theory holds, the Sun was once much fainter than it is today, Earth's oceans should have remained frozen until perhaps two billion years ago, thawing as the Sun grew brighter. Instead, sedimentary rocks show that at least some parts of the oceans have been liquid since their formation more than three billion years ago, suggesting that Earth's early climate was temperate. Somehow, the world adjusted to match the pace of solar change. The thermostat for early Earth seems to have been a concentration of carbon dioxide in the atmosphere perhaps 200 times greater than today's and a much tighter seal on the planet's energy supply. Then carbon dioxide levels diminished as the Sun brightened, rendering the atmosphere more transparent to escaping heat energy. Like its namesake, the greenhouse effect sustained life that would have perished outside its protection.

As with El Niño, the term "greenhouse effect" has fallen on hard times; what once seemed a natural gift now looks like a deadly enemy. This ambivalence is nothing new. In 1827, for example, French physicist and mathematician Jean-Baptiste Fourier first noted that the heat-trapping action of the atmosphere helps warm the planet. Without understanding the role played by carbon dioxide, he also suggested that human activity could modify the insulating effect and alter climate.

Later, in 1896, the noted Swedish chemist Svante Arrhenius worked out a more complete theory of the greenhouse effect. By then, the ability of the carbon dioxide molecule to absorb infrared radiation was well understood. More carbon dioxide in the atmosphere, Arrhenius pointed out, meant greater retention of the heat radiated by the planet, and a warmer climate. Arrhenius was able to calculate that a doubling of atmospheric carbon dioxide would warm the globe by about nine degrees Fahrenheit. He also proposed that a reduction in atmospheric carbon dioxide contributed to the onset of glaciation. Both ideas are now increasingly accepted by climate scientists.

At the end of the nineteenth century, there was little concern that carbon dioxide levels might be increasing, and no proof that they were. Whatever was put into the atmosphere, most investigators believed, would be rinsed away as the air interacted with the sea.

The first inkling that this might not be the case came in the 1930s, when some researchers suggested that an observed sixty-year warming trend might have been triggered by the rising use of fossil fuels. Again, there was no objective proof that fuel burning actually increased the amount of carbon dioxide in the atmosphere. Not until the late 1950s did careful, systematic measurement of the gas begin. In 1955, Caltech graduate student Charles David Keeling met with Harry Wexler, a pioneering research meteorologist with the U.S. Weather Bureau, to lay plans for a carbon dioxide monitoring program. Working from the Scripps Institution of Oceanography, Keeling adapted existing recorders developed by the mining industry to measure the trace amounts of the gas in the atmosphere. In 1958, after preliminary tests in the Antarctic, he set up his instruments at a station above the 11,000-foot mark on Mauna Loa, a volcano on the island of Hawaii. There, facing the winds sweeping eastward over the open Pacific Ocean, the first continuous record of a greenhouse gas was slowly assembled.

The atmospheric concentration of carbon dioxide rises and falls in a regular pattern that follows the seasons. In the northern spring and summer, when most of the planet's land plants revive and spread, the gas decreases, taken up in photosynthesis. In autumn and winter, as the vitality of northern plants begins to wane, carbon dioxide levels rise. This seasonal variation is superimposed on the steady trends of any global change in average concentrations of the gas. In 1958, Keeling recorded an average concentration of 315 parts per million, an increase of 14 percent over an estimated 275 parts per million at the start of the Industrial Revolution more than a century and a half earlier.

As illustrated by smoke and haze hanging over São Paulo, Brazil, the burning of fossil fuels such as coal and oil has significantly raised atmospheric levels of carbon dioxide and other gases that trap heat radiating from the planet's surface. Increases in these "greenhouse" gases may promote global warming.

110

Since 1958, this value has marched steadily higher, reaching nearly 350 parts per million in 1988, a 13-percent jump in just thirty years. The Mauna Loa measurements proved beyond a doubt that the composition of the atmosphere is changing and that the rate of change has begun to accelerate. Unanswered questions remain, however. For example, at 350 parts per million, the increased carbon dioxide represents only about half of what has been produced. Scientists assumed for many years that the other half was being taken up by the oceans and green plants. But more recently researchers have begun to argue that plants are a source, not a sink, for carbon dioxide—that is, they generate more than they take up. Clearly, the carbon budget is not as neatly balanced as had been expected.

Despite some measure of uncertainty about the process itself, researchers are sure of the source of the increased levels of carbon dioxide: They have come largely from a huge and growing human appetite for the energy locked in Earth's great reservoirs of carbon.

SMOKE AND FIRE

By the late 1980s, world energy consumption had reached nearly nine billion tons of coal—or its equivalent in other fossil fuels—every year, with energy use doubling about every two decades. The chemistry of combustion is inexorable. Fossil fuels are primarily carbon and hydrogen. When they burn, these atoms combine with oxygen from the air to form carbon dioxide (or sometimes carbon monoxide) and water vapor. For every ton of coal consumed, more than a half ton of carbon enters the atmosphere as carbon dioxide. Oil, with more hydrogen and fewer carbon atoms in its composition, contributes about 30 percent less for an equivalent amount of energy; natural gas contributes only half as much as coal. Worldwide, the burning of fossil fuels emits about five billion tons of carbon into the atmosphere every year— roughly one ton for every human on Earth.

Not all carbon fuel is stored in fossil form. The world's forests constitute an immense reservoir. When trees decay, or when they are cut down and burned, they release carbon dioxide to the atmosphere. If new trees grow or are planted at the same rate that others are cut down, there is no net increase in atmospheric carbon dioxide: The new plants take up the excess. For centuries, however, more forest has been cleared than planted, a trend that has intensified in recent decades.

In the Amazon River basin, the cutting goes on at a furious pace, driven by large commercial ventures and by hundreds of thousands of settlers pouring into frontier areas on newly opened roads. They cut and clear, burn and plant. Scientists have counted as many as 6,000 fires blazing at once, adding to the great gaps in the forest that now scar the landscape. The Amazon is only one such site among many: Extensive fires are also burning in Africa, Central America, and the once-great forests of Southeast Asia. So rapidly are the tropical forests disappearing—an estimated fifty-four acres a minute, two million acres a month, an area the size of Bulgaria every

A million-square-mile pall of smoke covers the Amazon River basin as far as the Andes Mountains (upper left) in a 1988 photograph taken from the shuttle Discovery. Fires set to clear rain forest not only pump more than a billion tons of carbon dioxide into the atmosphere annually but also leave fewer trees to reabsorb it. The inset shows the clarity of the air in the same region in 1973, before slash-and-burn deforestation became widespread.

Drifting Saharan sand *(light and dark gray)* encroaches on a narrow band of irrigated farmland *(red)* along the Nile in the 1976 Landsat image at right. Streambeds snaking down from the mountains show no signs of vegetation and are bone-dry except during rare flash flooding, but the size of some of them indicates that rainfall was once more plentiful. The bright yellow patch in the lower left marks the edge of Egypt's western desert.

Desert engulfs the city of Timbuktu *(brown)* in western Africa during the nine years between the two Landsat images below. The upper picture, taken in 1976, shows Timbuktu surrounded by vegetation *(dark green)* fed by the Niger River flood plain at bottom and a network of waterways near the city *(blue)*. By 1985 *(lower image)*, water supplies have almost disappeared, and much of the growth has been replaced by sand *(pale green)*.

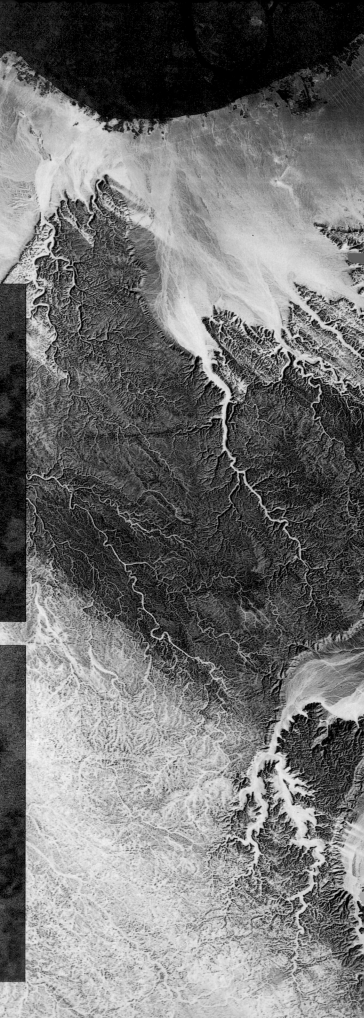

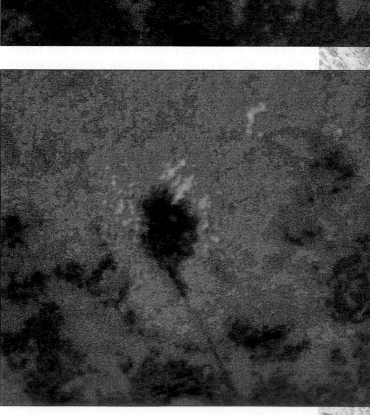

erodes readily before the force of the winds, adding to the shifting sand.

In much of the region, the process of desertification, as this is called, seems to feed upon itself and accelerate. Although scientists do not agree on the point, some see the key to desertification in the changing character of the surface and its effect on the radiation balance: Sand reflects more of the Sun's light and retains less moisture than does vegetation. These conditions may diminish the region's already limited rainfall, increasing the dryness of the climate and the growth of the desert. Thus it is possible that the Sahel peoples' age-old behavior has begun to alter the region's weather, exacerbating periodic droughts and the resultant famines.

THE FROZEN MIRROR

The spread of arid lands is not limited to northern Africa. Globally, an estimated 25,000 square miles of new desert form each year—in the Sahel, in northwestern India, in the northeastern corner of Brazil, and in the far northern and northwestern reaches of China. All of these seem to extend primarily from centers of human activity.

While replacing forests with sand increases Earth's albedo and may produce some general cooling, reduction of the planet's ice cover causes more heat to be absorbed by the climatic apparatus and average temperatures to rise. Ice reflects sunlight well and cools the air above it. When the ice sheets of higher latitudes creep toward the equator, the reflecting surface expands, causing further cooling and fostering the formation of more ice. At the moment, however, glaciers all over the planet either are retreating or have slowed their advance. Satellite observations show that several huge pieces of the Antarctic ice sheet have broken off into the sea. Overall, the evidence indicates perhaps a six-percent reduction in the north and south polar sea ice in the last decade. The smaller the frozen mirror, the more sunlight that is absorbed by the underlying land or water, and the greater the warming effect.

Dust in the atmosphere can also modulate a planet's albedo. On Mars, huge dust storms sometimes obscure the entire planet for days at a time, creating a reflective shroud that shields the surface from the Sun. On Earth, dust storms are far less spectacular, but when ash from an explosive volcanic eruption reaches the stratosphere, where turbulence is low and rainfall nonexistent, the spreading plume can have a persistent effect. For example, the 1815 explosion of the volcano Tambora in the East Indies violently ejected some fifty cubic miles of dust and gases into the upper atmosphere, blocking sunlight and cooling the surface by as much as two degrees Fahrenheit—enough of a chill that in the United States snow fell during July and August and newspapers from New England to the Carolinas called 1816 the "year without a summer."

But the effect of volcanic dust is transient, lasting at most a few years before the atmosphere rinses out the ashy debris. Decades, even centuries, can pass between major volcanic events, giving the planet time to recover its equilibrium. In the 1980s, scientists identified a more sustained atmospheric source of change in Earth's albedo: aerosols, systems of tiny particles dispersed in the air.

A couple of decades earlier, investigators studying the smog in Los Angeles during the late 1960s had analyzed the size of the particles in air samples. To their surprise, the grains were not randomly sized, as expected, but fell into two distinct groupings: fine particles with an average diameter of about a tenth of a micrometer (a micrometer is one millionth of a meter) and coarse particles a hundred times larger. Later studies showed those groups to be characteristic of airborne granules found all over the world. The two kinds of aerosol particles turned out to be chemically and qualitatively different. Coarse particles are typically dust—bits of soil or other material carried away by the wind and broken up into smaller and smaller pieces. However, the mechanical action of the wind cannot pulverize material into particles smaller than one micrometer. The finer particles proved to be micropellets of sulfuric acid. When a fuel containing sulfur is burned, one of the gases given off is sulfur dioxide, which forms sulfuric acid upon contact with water in the atmosphere. The longer sulfur dioxide remains in the atmosphere, the greater the proportion that is turned into fine sulfate particles, which can aggregate into particles ten times larger.

At first scientists thought the acidic motes were restricted to polluted areas of the world. But in 1978, investigators from the U.S. Environmental Protection Agency began studying the haze in an area of the Great Smoky Mountains National Park in Tennessee, presumed to come from trees in this pristine location. They found that most of the particles were sulfate aerosols. A year later, U.S. and Soviet scientists took air samples at a remote astronomical observatory in the mountains of Soviet Georgia, in a region known for its clear sky. Indeed, tests showed very few pollutants—but even here, the particles present were mostly sulfate aerosols.

Eventually, by comparing unique trace chemicals in different samples, researchers identified the source of the ubiquitous aerosols: coal-burning power plants or other sources of combustion, such as automobiles. Scientists now believe that most of the growing quantity of fine particulates in the atmosphere are sulfate aerosols. From the standpoint of the planet's thermal equilibrium, the floating particles are precisely the right size to scatter and disrupt sunlight, reducing the amount of solar radiation that reaches the ground and increasing Earth's albedo. Their acidity also tips a delicate balance in the planet's chemical climate.

LIFE'S ACID TEST

Signs of chemical change appear most vividly in a fluid almost synonymous with purity: rainwater. In many parts of the globe, particularly those downwind of heavily industrialized areas or large cities, rain is becoming more acidic, and the term "acid rain" has entered the popular vocabulary. Sulfate aerosols are a key player in this process, especially in areas far from sources of pollution. But sulfur is not the only chemical contributing to ac-

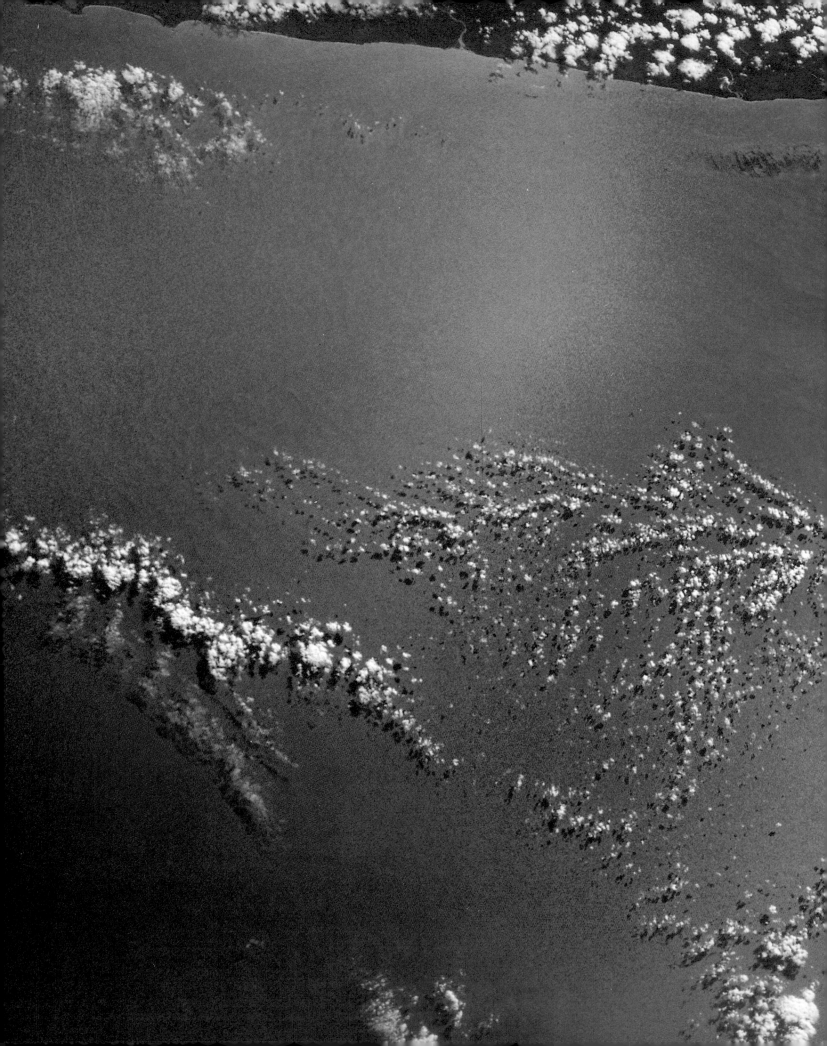

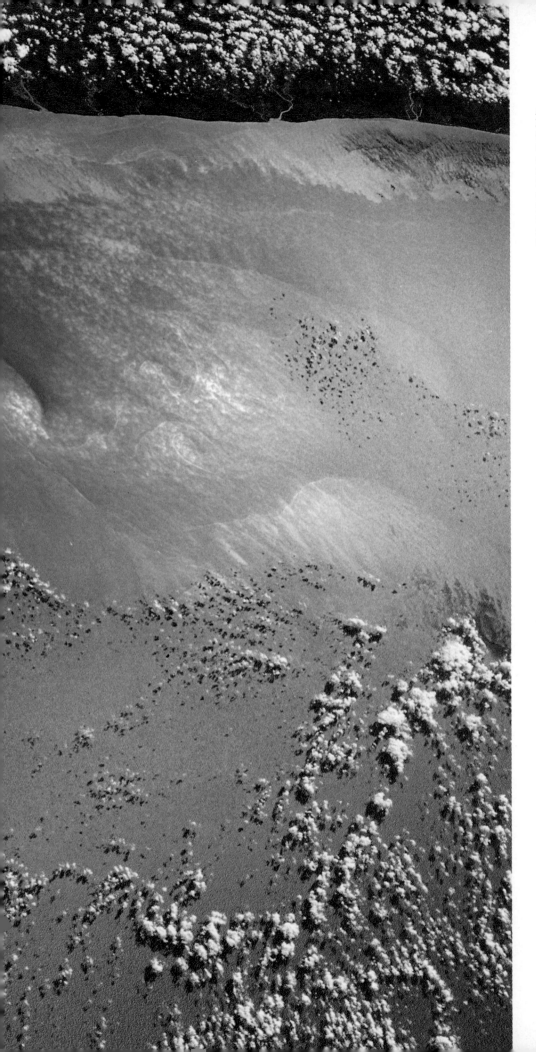

A golden orange glow over the
Indian Ocean denotes the presence
of atmospheric pollution, as
sunlight bounces off tiny particles
of carbon and other materials
resulting from industrial processes.
By increasing the reflectivity, or
albedo, of the planet, airborne
pollutants reduce the amount of
solar light and heat reaching the
surface, perhaps offsetting some of
the effects of greenhouse warming.

idification of the planet. Oxides of nitrogen produced in virtually all combustion, even of sulfur-free fuels, combine with moisture in the air to form nitric acid. Wherever one looks, the terrestrial trail of acid becomes clearer year by year.

For example, North Carolina motorists driving along the sinuous, narrow Blue Ridge Parkway between Asheville and Mount Mitchell might notice a few patches of dead trees along the scenic highway. The road ends in a parking lot a short way below the summit—the highest point in the eastern half of the United States. Here, whole stands of dead, gray trees are interspersed with those that are still green. The moribund evergreens of Mount Mitchell were healthy as late as 1983. Now the devastation extends well down the flanks of the mountain. It is as if a great plague suddenly descended on the forest. The same plague has touched much of the Appalachian chain that stretches from the Carolinas to New England. The maple trees in Canada's Quebec province, the pine groves in the American southeast and in parts of California—all of them show signs of damage.

In Germany, a similar phenomenon is called *Waldsterben*—forest death. Scientists noticed it in the Black Forest in the early 1970s. In 1984, they reported that half of the German forests had been harmed. Now trees of many species covering more than 17 million acres, from Switzerland to Sweden, are dying or showing signs of injury.

Sickly trees are only one symptom. In the high mountains of New York's Adirondacks and throughout Scandinavia, thousands of lakes and streams have become sterile, devoid of fish or other biota.

THE PROCESS OF DESTRUCTION

Almost certainly, the addition of sulfates and nitrous oxides to the lower atmosphere is affecting trees, soils, and lakes. But precisely how these compounds do their destructive work is still not completely understood. For example, the mechanisms may be different in each area. In cloud-bound regions, the acids in droplets of moisture may harm trees directly. In other areas, acids accumulate in lakes where, if not chemically neutralized by minerals in the lake bed, they become strong enough to kill algae and fish. In some soils, the acid rain leaches out nutrients, starving the trees of food and making them more vulnerable to cold, disease, or insect pests; in others, the acids may mobilize chemicals, such as aluminum, that are toxic to plants. Scientists have even suggested that acid rain stimulates the growth of mosses that may kill roots.

Ozone, a major component of urban pollution, also seems to contribute to the killing off of lakes and forests. The unstable assemblage of three oxygen atoms is toxic—so much so that it is used as an industrial germicide. Yet the compound exemplifies one of the many ironies of climatic change. As harmful as it is to life on the surface of the planet, ozone serves a benign purpose some fifteen miles overhead. High in the stratosphere, oxygen molecules are split asunder by solar radiation. In the process of recombining into three-atom molecules, the oxygen absorbs biologically harmful ultraviolet wavelengths, providing a stratospheric shield for creatures on the surface. In that arena, ozone now appears to be in trouble.

In the spring of 1985, British scientists working in Antarctica reported a disturbing discovery: Ozone levels over Halley Bay had declined 40 percent between 1977 and 1984. Other scientists soon confirmed the report and established that the ozone hole, as the thinning soon came to be called, extended over the entire Antarctic continent and beyond. Atmospheric chemists had begun to worry about possible ozone destruction by human activities in the early 1970s, but the depletion measured over the South Pole exceeded their direst predictions.

Because ozone is present in the stratosphere in concentrations of less than one part per million parts of air, relatively small changes may produce large effects. A decline in ozone would expose the Earth's surface to more ultraviolet rays, which are so energetic that they can break apart organic molecules—harming crops, causing skin cancer in humans, and conceivably dam-

Long tendrils of underwater lobelia rise from a dense mat of algae toward the surface of an acidified lake in Sweden. Sulfur dioxide emissions deposited by rainfall have altered the lake's biology, encouraging the growth of acid-loving plants but killing off most fish. The disappearance of microscopic organisms and phytoplankton gives the water an exceptional clarity.

Tracing Ozone Loss

Outlined below is one theory for the springtime depletion of ozone over Antarctica. The process depends both on chemical products of chlorofluorocarbon (CFC) breakdown and on unique atmospheric conditions. Extreme cold during the Antarctic winter and a vortex of winds at the South Pole help form clouds in the stratosphere, where ozone resides. Ice crystals in these clouds serve as a base for chemical reactions that release chlorine, one of the leading instruments of ozone destruction.

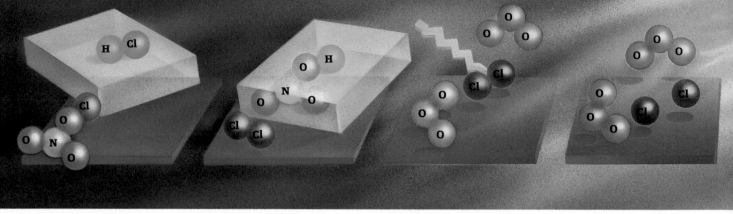

Polar night. Ice crystals *(transparent block)* formed in the stratosphere during the perpetual dark of Antarctic winter interact with two compounds resulting from the natural breakdown of CFCs: hydrogen chloride (HCl), which diffuses within the crystal, and chlorine nitrate (ClNO₃), which attaches to its surface.

Within milliseconds, the chlorine nitrate reacts with the hydrogen chloride to form a molecule of chlorine (Cl₂) and a molecule of nitric acid (HNO₃). The nitric acid remains trapped in the ice crystal, but the chlorine molecule goes free.

Polar day. The reappearance of the Sun during the Antarctic spring sets into motion the destruction of ozone molecules, which consist of three atoms of oxygen. The process begins when a molecule of chlorine is struck by a photon *(lightning bolt).*

A photochemical reaction triggered by the photon breaks the chlorine's molecular bond, unleashing two atoms of chlorine.

aging the tiny, one-celled phytoplankton at the base of the marine food chain. Moreover, the loss of the high-altitude ozone would lessen the stratospheric warming contributed by this greenhouse gas, causing the high atmosphere to cool, with unpredictable climatic effects.

At first, the South Pole data indicated that the ozone hole was short-lived, appearing only in the austral spring; during the Antarctic summer, ozone concentrations returned to normal levels. But in 1986, the thinning became more pronounced, suggesting that more and more ozone was being destroyed. The suspected agents of destruction were the chlorine-containing industrial chemicals called chlorofluorocarbons. Once these compounds reach the stratosphere, harsh solar ultraviolet light wrenches them apart, liberating chlorine, which can catalyze or speed the destruction of ozone *(above).*

With concern growing, NASA organized an expedition in August 1987 involving 150 scientists and support personnel to conduct the Airborne Antarctic Ozone Experiment. Working from a base in Punta Arenas, Chile, the scientists flew repeated missions on two heavily instrumented aircraft: a modified DC-8 passenger plane and a high-flying ER-2, cousin to the U-2 spy plane. By sampling the atmosphere at different levels and comparing the

Each of the chlorine atoms strips an atom of oxygen from the ozone molecules, thereby creating two oxygen molecules (O_2) and two chlorine monoxide (ClO) molecular fragments.

The chlorine monoxide fragments combine to form what is known as a chlorine monoxide dimer (Cl_2O_2). Sunlight again comes into play as a photon then strikes the dimer.

The photon causes the chlorine monoxide dimer to break down into a molecule of chlorine dioxide (ClO_2) and a single chlorine atom.

Because it is molecularly unstable, the chlorine dioxide also breaks down, forming an oxygen molecule and another lone chlorine atom. The two chlorine atoms released in this and the previous stage are once again ready to attack ozone. The last four steps of the cycle will repeat throughout the Antarctic spring.

results with data gathered on the ground, the team hoped to isolate what was really going on.

Even as the expedition was at work, however, political leaders were responding to the threat of ozone depletion. In September, representatives of some twenty-three nations met in Montreal to forge an agreement that would restrict the production of chlorofluorocarbons and eventually begin to phase them out. The resulting treaty marked the first international legislation aimed at preventing a global climate change. The preliminary findings from Antarctica, meanwhile, led to proposals to make the terms of this precedent-setting agreement even stronger.

By March 1988, the results from the Antarctic expedition were in. The ozone hole was almost certainly caused by industrial sources of chlorine. But the mechanism was not what researchers had expected. Apparently, the chlorine's ozone-destroying ability was being amplified by ice crystals in thin clouds that form under extremely cold conditions in the polar stratosphere. The key chemical reactions began on the surface of ice particles in the clouds and were later energized by solar radiation, but only while a stable pattern of winter winds circled the South Pole, keeping out ozone from other parts of the atmosphere.

Two weeks after scientists presented their South Pole ozone analysis, the DuPont corporation, the world's largest manufacturer of chlorofluorocarbons, announced plans to phase out their production. Even so, concentrations of these inert materials in the atmosphere are expected to continue to climb—and ozone levels to decline—for decades to come.

New satellite data showed a decline over the last seventeen years of up to three percent in stratospheric ozone quantities above the United States, Europe, and the Soviet Union. Early in 1989, a second expedition, this one flying from Norway, found both chlorine and stratospheric clouds over the Arctic as well. The ozone layer above the North Pole was chemically primed for destruction, if the appropriate conditions were to occur.

Perhaps the most striking aspect of the chlorofluorocarbon story concerns the rapidity of the climate changes these chemicals induced. In less than forty years of widespread use, chlorofluorocarbons have altered the chemical composition of the atmosphere sufficiently to become a significant contributor to the greenhouse warming of the lower atmosphere and to cause measurable losses of ozone in the stratosphere.

ALCHEMY AT WORK

Carbon dioxide. Methane. Nitrous oxides. Chlorofluorocarbons. That these gases, leaking into the atmosphere in trace amounts, could trigger a global warming and transform the climate seems almost an alchemist's fantasy. Yet scientists are increasingly convinced of its reality. They cannot predict in detail what climate change will bring, or exactly how quickly the changes will occur. The ocean's ability to absorb carbon dioxide is uncertain, as are the detailed chemistry of the ozone layer and the precise ways acid rain damages a forest. Nevertheless, a radically different climate may be on the horizon. Already the available evidence suggests that average global temperatures have climbed about one degree Fahrenheit over the last hun-

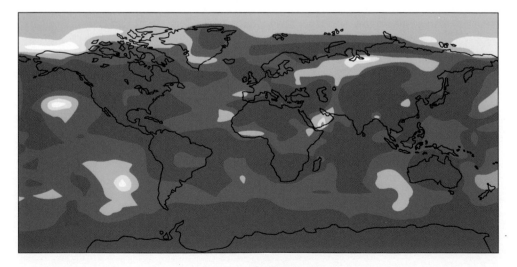

A computer simulation of Earth's climate, from the Goddard Institute for Space Studies, produced this color-coded map showing temperature changes that would occur during June, July, and August of 2050 if greenhouse gases were to double by 2030 from 1958 levels. Intense warming—of as much as nine degrees Fahrenheit *(brown)*—would occur over landmasses, including Antarctica, and over oceans near the equator. Rust and gold denote moderate increases, yellow and white little or no increase, and blue a slight cooling.

dred years. By the end of the next century, most scientists now believe, the Third Planet is likely to be significantly warmer—warmer than it has been for a million years.

CLIMATE BY THE NUMBERS

Predicting these futures, like forecasting weather and short-term climatic changes such as El Niño, has become the province of numerical models of the atmosphere and ocean, the so-called coupled models that are constructed in the world's most powerful computers. Climate researchers are keenly aware that their models do not reflect all the possible climate-related interactions—between temperature and humidity, for example, or between carbon dioxide and trees. As a starting point, then, they run their models to simulate present and past climates, looking for important connections in geophysical histories. And they try to improvise ways of including the effects of other factors, such as individual clouds, that are too small to specify in the simulation. Much of the subtle natural detail must be rendered as a mathematical generalization, however. For example, the interlinked responses of vegetation and climate to increased carbon dioxide are a circular puzzle. Plants might flourish and take up more carbon dioxide, slowing the warming. But warmer temperatures would cause dead vegetation to decay more quickly, releasing more carbon dioxide into the air and intensifying the warming. Similarly, models cannot accurately depict the oceans' ability to absorb carbon dioxide, a severe handicap since the oceans hold more of the gas than does the atmosphere. The problem is that most models compute the climate as it would be at equilibrium, after changes have occurred and the environment has settled down. This ignores the fact that the moving parts of the models are running at different rates, making them hard to mesh. Greenhouse gases, for example, increase their effect on a time scale of decades, while it takes the ocean a thousand

A similar computer climate model here predicts changes in precipitation. More frequent and severe droughts would occur in the low and middle latitudes *(brown and rust),* while precipitation would increase *(light and dark blue)* nearer the poles and in regions where trade winds blowing from the Northern and Southern Hemispheres converge. White indicates little or no anticipated change.

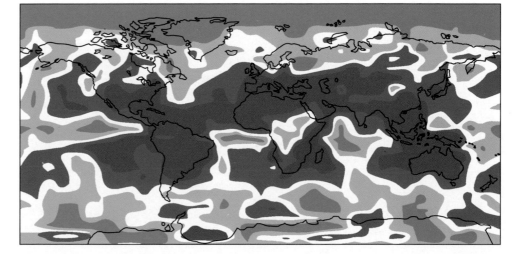

years or more to reach an altered state of equilibrium. Reconciling such disparate processes remains a major challenge for atmospheric scientists.

SCENARIOS

Despite their drawbacks, climate models are persuading many scientists that major changes are on the way, although there is no real consensus on the type, amount, or rate. Virtually all general circulation models show an increase in global temperature of three and one-half to nine degrees Fahrenheit over the next century if greenhouse gas emissions continue and effectively double current levels of carbon dioxide. In fact, some scientists believe that even if all emissions ceased today, enough carbon dioxide has already been added to the atmosphere to ensure future warming.

Calculations published in 1988 by James Hansen and his coworkers at NASA's Goddard Institute for Space Studies in New York suggest that average global temperatures may equal the highest attained during the current or past interglacial period, even if drastic curtailment of greenhouse gas emissions could halt additions in the atmosphere by the year 2000. Still higher temperatures will occur, according to the model, if gas levels continue to increase. NASA researchers think the warming will be apparent first in tropical oceans, China, and central Asia, and at both poles. Moreover, they say, warmer oceans may spawn more hurricanes, and the different heating rates of land and water could create giant windstorms over the continents.

Their findings agree only in part with calculations based on a very different model developed by Sukuro Manabe, Kirk Bryan, and their colleagues at the Geophysical Fluid Dynamics Laboratory. These modelers also find a global temperature rise, but regional patterns in the warming differ sharply from Hansen's projections. The Princeton model suggests, for example, that temperatures would not significantly rise near Antarctica because of the effects of the strong circumpolar ocean current. In fact, their calculations predict

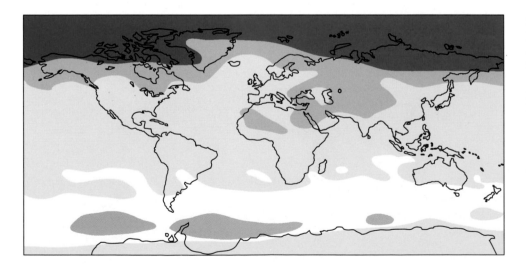

A climate model by the Geophysical Fluid Dynamics Laboratory (GFDL) traces annual mean temperature change stemming from a quadrupling of 1950 atmospheric carbon dioxide levels. Results differ markedly from north to south. High northern latitudes witness pronounced warming *(red)*—up to eighteen degrees Fahrenheit—while comparable southern regions experience little or no change *(white)* or cooling *(lavender)*. Orange and yellow denote smaller increases.

that the Northern Hemisphere would warm much more than the high latitudes of the Southern Hemisphere, where there is far more water than land and where currents and upwelling of cool, deep water seem to be more important in moderating temperature increases. The land and ice-covered regions of the Northern Hemisphere lack such a buffer; the largest temperature increases predicted by the Princeton model would occur near the North Pole.

Disparities from model to model also occur for such factors as global rainfall and soil moisture. A model developed by Manabe shows that more greenhouse gases reduce soil moisture in midcontinent areas such as the North American Great Plains, western Europe, northern Canada, and Siberia. Other studies show that the greatest effects, in the United States at least, will likely come in the arid southwest, where even small decreases in precipitation would have major effects on farming and ranching. Still other studies, based on different models, suggest a drought in the southeastern United States. At the same time, many models call for increased precipitation near the equator, where monsoon rains provide the major source of water. In fact, the various scenarios—scientists do not call them forecasts—of precipitation changes, laid side by side, show a host of both major and subtle differences, especially at a regional level. What is but a blip on a global basis—a new dust bowl in the U.S. grain belt, for example, or the failure of the East African rains that fill the Nile River—could mean disaster for farmers and social unrest for whole countries, but the science of modeling climate change is not advanced enough to be a reliable guide for government action.

If scientists cannot yet read the geophysical future in detail, they are able now to create a rough picture of what a warmer Earth would be like. As the oceans warm, for example, seawater will expand, raising sea levels slightly. If glaciers and pack ice begin to melt, the high-water line could rise several feet by the end of the next century. If warming continues, the great western Antarctic ice sheet might eventually free itself from the continent and plunge

This map—also generated by a GFDL computer model—shows changes in soil moisture levels during the summer months June, July, and August resulting from a doubling of atmospheric carbon dioxide. Effects range from an increase of more than 10 percent to a reduction of greater than 20 percent, as indicated respectively by lavender, white, yellow, orange, and red. Results were not simulated where land is permanently covered by ice *(gray)*.

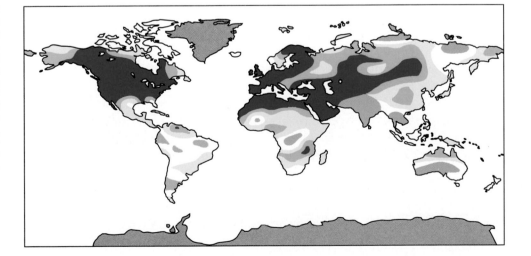

into the ocean, raising the sea's surface by some twenty feet and greatly altering the coastlines of the continents.

More speculative, perhaps, are estimates of the impact of a global warming on plant and animal life. The projected changes are expected to be large and swift enough to cause enormous dislocations across the planet's surface. Biological communities may be disrupted as their habitats shift and they are forced to follow their food supplies. Some—perhaps many—species will become extinct. The fragile Arctic tundra may disappear. In the Northern Hemisphere, forests could die off at their southern extremes but expand only slowly northward, leaving vast stretches of meadowland in their wake. Agriculture might have to take its grain belts northward.

ADAPTING TO A WARMER EARTH

From a plant's point of view, the scenarios of warming are not all bad. According to one plant scientist, growth rates will increase as Earth moves from being carbon-starved to carbon-fertilized. Many animals will also respond favorably to a warmer planet. Locusts, aphids, and moths become more active and breed more rapidly as temperature and humidity rise, and will undoubtedly follow the tropical climate as it creeps north. Higher temperatures will induce some kinds of reptiles to produce more females than males, perhaps fueling a population explosion among alligators and the like. Parasites and diseases now climatically insulated in the tropics will expand northward and southward from the equator.

Although some of these changes seem almost apocalyptic, scientists note that even the most dramatic climatic shifts are very small compared to some past terrestrial calamities—alterations that fundamentally altered Earth's environment without extinguishing the tough, vital spark of life.

The first such disaster was self-administered by the planet's early population. When blue-green algae acquired a new and more efficient way to feed themselves through photosynthesis, they began to produce a lethal waste product—oxygen—which at the time was present in Earth's atmosphere only in trace quantities, if at all. As the photosynthetic algae released the reactive, toxic gas into the atmosphere, they shaped the first biologically induced climatic change. By creating an oxygen-rich atmosphere, they finally poisoned themselves and their relatives with their own exhalations, exiling most early forms of life to anaerobic niches in swamps and soils. Those species able to adapt became the ancestors of almost all who live on Earth today.

Another great disaster came from outside, administered from the heavens. Collisions between Earth and other heavenly bodies were frequent early in the planet's history. Many scientists now believe that one such collision 66 million years ago set off a dramatic sequence of environmental upheavals that wiped out most flowering plants, millions of marine species, and virtually every terrestrial animal too large to find shelter—including the dinosaurs. The course of events leading to the largest mass extinction in the fossil record probably began when a comet or asteroid perhaps six miles across plummeted through the atmosphere. According to planetary geologist Eugene Shoemaker, with the U.S. Geological Survey in Flagstaff, Arizona, the comet broke into at least two huge chunks, one crashing to the ground near Manson, Iowa, and the other diving into the Atlantic Ocean. The impact released energy equivalent to a hundred million one-megaton bombs, enough to ignite nitrogen in the atmosphere and form a worldwide cloud of toxic smog; it also ejected huge quantities of dust and hundreds of cubic miles of water into the upper atmosphere. The smog, dust, and water combined to shroud the world in near total blackness.

The darkness may have persisted for several months before the cloud dissipated, in the meantime killing plants and chilling the globe. But the worst was yet to come. The water in the atmosphere would have reacted chemically with the smog, making the rainfall as corrosive as battery acid. Falling on the land, it would have killed whole forests; in the oceans, it would have dissolved the very shells and exterior skeletons of many marine organisms. Lightning subsequently kindled the dead forests into firestorms of an unprecedented scale. These would have blackened the sky and again lowered the temperature—geologic deposits around the world show that a pronounced layer of soot was laid down at about this time. Any creature that survived the chill, the fires, the acid rain, and the lack of growing food would then have faced yet another crisis. When the skies cleared, the sharp rise in carbon dioxide from burning most of the world's biomass would have quickly raised temperatures perhaps eighteen degrees Fahrenheit, where they remained for a hundred thousand years.

Some may find a cautionary tale in the story of blue-green algae fouling their own nest. Others may see a lesson of equal force in rocky messengers from the heavens that—like the biblical flood—extinguished the most advanced forms of life on the planet, along with many lesser forms. Given the history of Earth's bombardment from space, that, too, could happen again, regardless of human actions.

Despite both of these catastrophes, however, life survived. It adapted, new forms evolved, and the biogeochemical cycles endured; the climate changed, and changed again. Somehow the complicated apparatus of Earth kept the alterations within tolerable bounds. The planet has been neither scorched by a runaway greenhouse effect nor smothered in a sheath of ice, nor is there any convincing evidence that either extreme is likely for hundreds of millions, perhaps billions, of years to come.

The climate of the Third Planet will continue to evolve and adjust, and just as certainly, some kinds of living things will keep adapting to those changes. But for any given species, even a dominant one, there are no guarantees—as the extraordinarily successful and adaptable dinosaurs ultimately discovered. Where one species becomes extinct, another is likely to find a temporary niche, or even an evolutionary advantage. For as long as this world wears the marbled skin of ocean and atmosphere that is its unique signature, it will remain the Solar System's garden of life.

Dawn over the Atlantic reveals a
rich tapestry of color: golden
sunlight on the water, dark gray
masses and silvery wisps of clouds,
the thin blue veil of the upper
atmosphere, and the deep black of
space. The dynamic interplay of
Earth's oceans, landmasses, and
atmosphere has fostered a climate
of remarkable resilience, able to
adapt to continual changes and
preserve the delicate balance of
forces that sustain life.

GLOSSARY

Accretion phase: the stage early in the evolution of a planet or star when its constituent matter is coalescing to form larger bodies.

Acidification: a process that tips the natural chemical balance toward higher proportions of acids. In Earth's atmosphere, nitric and sulfuric acids formed by chemical reactions are returned to the surface as acid rain and as a dry fallout of acidic particles.

Aerosol: tiny particles suspended in a gas. Aerosols in Earth's atmosphere include particles of dust, volcanic ash, sea salt, combustion products, and chemical pollutants.

Albedo: the ratio of solar radiation reflected by a surface to the radiation received by it, usually given as a percentage. For example, Earth's planetary albedo is about 34 percent.

Alpha particles: helium nuclei consisting of two protons and two neutrons given off by one form of radioactive decay.

Aphelion: the point in the orbit of a planet or a comet where it is farthest from the Sun.

Asthenosphere: a layer of mantle rock that lies between the lithosphere and the mesosphere. Because asthenospheric rock is relatively viscous, it is believed to provide a yielding surface on which lithospheric plates can glide.

Atmosphere: a gaseous shell surrounding a planet or other body.

Biogeochemical cycles: interacting processes that circulate nutrient chemicals such as carbon, nitrogen, and phosphorus through the air, sea, solid Earth, and biomass.

Biosphere: the totality of a planet's living things and their habitats; on Earth, a zone reaching from a few miles within the planet into the upper atmosphere.

Carbon cycle: the biogeochemical cycle that circulates carbon through Earth's environment. The carbon cycle is a key regulator of terrestrial climate.

Catastrophism: the theory that the present Earth and its creatures are the end result of a series of catastrophes larger than those experienced today, such as the biblical flood of Noah.

Celsius: a scientific temperature scale in which 0 degrees is the freezing point and 100 degrees the boiling point of water.

Chlorofluorocarbons (CFCs): volatile, inert chemical compounds containing carbon, chlorine, fluorine, and hydrogen, often used as spray-can propellants and refrigerants. In the stratosphere, the interaction of CFCs and ultraviolet radiation releases chlorine atoms in a process that accelerates the destruction of ozone, weakening the planet's shield against biologically dangerous radiation.

Continental drift: the theory that the continents are not static masses of land but part of giant, moving plates that make up Earth's crust.

Convection: the transfer of heat in a fluid or gas by the movement of currents from hotter to cooler regions.

Core: the innermost region of a body. Earth's core is thought to consist of two irregular, concentric spheres: an outer core of molten metal and a solid inner core believed to be mostly iron.

Crust: the brittle outermost layer of Earth, from which the continents and ocean floors are formed. Under the oceans, the crust is about five miles thick; under the continents, it is about five times thicker.

Daughter element: the element into which radioactive decay transmutes a parent element. For example, radium is a daughter of uranium.

Deforestation: the large-scale destruction of forests.

Desertification: the spread of desert into once-fertile areas.

Earthquake: the release of energy in the form of a shock wave built up by the differential movement of crustal blocks along a fault.

Easterlies: systems of prevailing, westward-blowing winds. Tropical easterlies are often called the trade winds for their role in propelling sailing ships across the oceans.

Eccentricity: the degree to which an ellipse deviates from a circular shape.

El Niño: a reversal of Pacific oceanic and atmospheric conditions that weakens easterly trade winds, bringing heavy rains to the normally dry coast of South America, diminishing monsoon winds and rains over the Indian Ocean, and causing drought in equatorial Africa. Because the phenomenon occurs near Christmas every few years, Peruvians named it El Niño—Spanish for "the child"—for the infant Jesus. *See* La Niña.

Epicenter: a point on Earth's surface directly above the subterranean focus, or source, of an earthquake.

Fault: a region of weakness in Earth's crust. Differential motion on either side of faults builds up enormous tension; when the fault fails, the pent-up energy is released, causing an earthquake.

Fold: a geologic formation caused by compression or a buckling of the crust.

Foraminifera: single-cell sea animals that have microscopic shells composed of calcite.

Fossil: the remains, trace, or impression of an ancient plant or animal preserved in sedimentary rock.

Front: the transition zone between air masses of different densities and temperatures.

Gaia theory: a hypothesis proposing that Earth is a self-regulating system, in which living organisms interact with the physical world to produce environmental conditions suitable for life.

General circulation model: a mathematical model used to simulate conditions in the global atmosphere, which the model simplifies into a series of horizontal layers divided into a grid. Quantities such as temperature, winds, water content, and pressure are added for each grid; the model then predicts how these quantities will change over time, following known physical laws.

Geophysics: the study of physical phenomena that act on or originate within the Earth.

Global radiation balance: the balance between the energy received by Earth from the Sun and the energy the planet radiates back into space.

Gondwana: the name given by German meteorologist Alfred Wegener to a hypothetical Mesozoic Era landmass that included the continents that are now Africa, South America, Australia, and Antarctica.

Greenhouse effect: a phenomenon in which radiation is selectively transmitted and absorbed by gases in an atmosphere, admitting incoming short-wavelength solar radiation but blocking outgoing long-wavelength infrared, thus trapping heat near the surface.

Greenhouse gas: atmospheric constituent—such as carbon dioxide, nitrogen oxides, water vapor, chlorofluorocarbons, or methane—that warms the atmosphere by trapping outbound infrared radiation.

Guyot: a flat-topped seamount that formed as a volcanic cone and then was submerged beneath the ocean.

Gyre: ringlike movement of ocean currents, turning clockwise in the Northern Hemisphere and counterclockwise in the Southern Hemisphere. The Gulf Stream is part of a large gyre carrying water around the Atlantic Ocean basin.

Hadley cell: in a planetary atmosphere, a convection cycle that pushes gases rising from warm areas toward cooler ones; the cooled gases then sink and return to the warm area to begin the cycle again.

Half-life: the time required for half the atoms in a sample of radioactive material to decay into a daughter element. The half-life of uranium-238, for example, is 4.5 billion years.

Hot spot: an area where rising magma plumes burn holes through Earth's crust; thought to be the origin of some volcanoes.

Ice age: a period during which polar ice sheets extend farther toward the equator, sometimes covering much of the planet's land areas with glaciers.

Interglacial: relatively short periods during an ice age when glaciers retreat and temperatures rise.

Isotope: one of two or more forms of a chemical element that have the same number of protons but a different number of neutrons in the nucleus.

La Niña: air-sea conditions in which Pacific surface waters are unusually cool, trade winds are strong, and in Asia, monsoons bring heavy rains. The term—coined to identify an opposite and alternating set of conditions called El Niño—means "the girl" in Spanish.

Lithosphere: the rigid outer layer of the Earth above the asthenosphere that includes the crust and the upper part of the mantle. The lithosphere is divided into large plates that, propelled by convective currents in the mantle, slide on the viscous layer of the asthenosphere.

Magma: molten material formed deep within the Earth that may force its way through the crust to the surface, where it cools as lava.

Magnetic field: the area around a magnet, an electrical current, or a charged particle in which a magnetic influence is felt by other currents, fields, and particles. Planetary magnetic fields, like those of a simple bar magnet, exhibit north and south poles linked by lines of varying magnetic strength and direction.

Magnetometer: a device for measuring the strength and direction of a magnetic field.

Magnetosphere: a large, energetic envelope of magnetic force lines shaped by interactions between a planetary magnetic field and the solar wind, the flow of particles and radiation from the Sun.

Mantle: the layer of the Earth that extends from beneath the crust approximately 1,800 miles to the core.

Mathematical models: a series of mathematical equations used to simulate processes and relationships in the real world; also called numerical models.

Mesosphere: from the Greek *mesos* for "middle," a term used to denote an intermediate layer in both the atmosphere and the Earth's mantle. In the atmosphere, the mesosphere occurs between thirty and fifty miles above the surface; in the mantle, it lies between the core and the asthenosphere and measures 1,550 miles from top to bottom.

Meteorite: a small metallic or rocky body fallen to a planet's surface from space and believed to represent debris from the time of Solar System formation.

Mohorovičić discontinuity: a shallow layer approximately twenty-five miles below the surface under continents and five miles below the ocean floor where seismic-wave velocity changes. Now used to denote the boundary between the planet's crust and mantle, it was named for its discoverer, the Yugoslavian geophysicist Andrija Mohorovičić, and is familiarly known as the Moho.

Ocean currents: persistent streams of moving ocean water, occurring at the surface and at depth. Surface currents are mainly driven by the wind and the planet's rotation and may affect only the upper few hundred feet of the sea. As they cool, the currents descend and flow through the deep ocean, driven mainly by differences in water density.

Oceanic ridge: a seam formed where material rising from the Earth's mantle flows into the deep ocean, creating new crust and causing the floor to spread outward from the ridge. A global system of midocean ridges some 34,000 miles long encircles the planet, broken by cracks, or faults, in the seabed.

Ophiolite: rock composed of both magma and oceanic sediments.

Ozone: the highly reactive, unstable three-atom form of the element oxygen, which normally occurs as a two-atom molecule. Near the surface, ozone is a toxic pollutant; in the stratosphere, it absorbs incoming solar ultraviolet rays, which are harmful to life on the surface.

Pangaea: the name given by German meteorologist Alfred Wegener to a hypothetical ancient landmass from which all modern continents were formed. *See* Gondwana.

Parent element: a material that transmutes into another element, called a daughter element, through radioactive decay. For example, the parent uranium decays into the daughter radium.

Perihelion: the point in the orbit of a planet or a comet where it is nearest the Sun.

Photosynthesis: the mechanism in plants that uses solar energy to make carbohydrates and other organic nutrients from carbon dioxide; in the process, carbon is taken up, or fixed, and oxygen is released.

Plate tectonics: the theory that the Earth's crust is a mosaic of approximately nine large moving plates and several smaller ones, which rest on the asthenosphere. Plate tectonics theory holds that processes deep in the planet propel the plates, causing continents to migrate and producing the structural tensions released in the form of earthquakes and volcanic eruptions.

Precession: the circular, wobbling motion of Earth's axis of rotation caused primarily by the gravitational attraction of the Moon. The circle is completed approximately every 23,000 years.

P-waves: primary seismic waves that move through the body of a planet, propagated like sound or other compression waves. *See* S-waves.

Radioactive decay: the spontaneous breakdown of an unstable nucleus by the release of subatomic particles and heat.

Radiolarians: single-cell marine organisms with an external skeleton composed of radiating spikes of silica.

Richter scale: a logarithmic scale, determined from seismic records, that relates the amplitude of earthquake waves received at a seismograph to the energy released by an earthquake. Theoretically open-ended, Richter magnitudes have a practical upper limit of about nine.

Rift: the scar created when a tectonic plate splits apart.

Seafloor spreading: the process by which the seafloor expands from spreading centers along oceanic ridges as magma wells up from the Earth's interior to create new crust.

Sedimentary rocks: rocks formed by the consolidation and compression of sediments deposited in layers.

Sediments: solid particles, deposited in a fluid, that form sedimentary rocks when compacted.

Seismic profiling: the technique of using sound waves from an explosion or other source to probe subterranean geologic structure; also called seismic prospecting by petroleum geologists.

Seismic velocity: the speed of propagation of seismic waves through media of different properties; velocities increase as material becomes denser.

Seismic waves: the form taken by energy released in an earthquake. Seismic waves travel either through the body of the planet or along the surface.

Seismogram: the record of seismic waves detected by a seismograph.

Seismograph: an instrument designed to detect, amplify, and record ground motion caused by passing seismic waves.

Seismology: the study of earthquakes and their relation to the planet's structure and dynamics.

Shock wave: a sudden discontinuity in the flow of a gas, liquid, or plasma characterized by increases in temperature, pressure, and velocity.

Subduction zone: the region where a spreading seafloor plate, for example, is forced under a continental one, or subducted. Subduction occurs mainly along the deep ocean trenches, where crustal material is driven back into the mantle.

Supernova: the explosive death of a massive star. The debris of such explosions is believed to serve as building material for new stars and their planetary systems.

S-waves: secondary, or shear, seismic waves that travel through the body of the planet, undulating at right angles to their direction of movement. *See P-waves.*

Thermocline: a distinct zone of ocean water where temperatures decline sharply with depth. It serves as a natural barrier to vertical mixing between the relatively light surface waters and the denser, colder water of the deep sea.

Thermodynamics: the study of how heat and mechanical energy are linked.

Thermonuclear fusion: a reaction between atoms that leads to their fusion into a heavier atom, releasing energy in the process. The Sun, for example, generates its energy by fusing hydrogen into atoms of helium.

Trace gases: gases that are present in very small amounts. In Earth's atmosphere, carbon dioxide, at 350 parts per million, is a trace gas.

Transform fault: a crack in the planet's crust induced by differential tensions along the spreading center of an ocean ridge. Transform faulting divides the terrestrial midocean ridge system into many segments.

Trench: a narrow depression in the seafloor associated with the subduction of an oceanic plate under an adjacent continental plate; trenches are the deepest parts of the oceans.

BIBLIOGRAPHY

Books

American Meteorological Society, *National Conference: Scientific Results of the First GARP Global Experiment.* Boston: American Meteorological Society, 1986.

Bach, Wilfrid, Jürgen Pankrath, and William Kellogg, eds., *Man's Impact on Climate.* Amsterdam: Elsevier Scientific Publishing, 1979.

Barth, Michael C., and James G. Titus, eds., *Greenhouse Effect and Sea Level Rise.* New York: Van Nostrand Reinhold, 1984.

Baugher, Joseph F., *The Space-Age Solar System.* New York: John Wiley & Sons, 1988.

Briggs, Peter, *Water: The Vital Essence.* New York: Harper & Row, 1967.

Broecker, Wallace S., *How to Build a Habitable Planet.* Palisades, N.Y.: Eldigio Press, 1985.

Calder, Nigel, *The Restless Earth.* New York: Penguin Books, 1979.

Cattermole, Peter, and Patrick Moore, *The Story of the Earth.* Cambridge: Cambridge University Press, 1985.

Chorlton, Windsor, and the Editors of Time-Life Books, *Ice Ages* (Planet Earth series). Alexandria, Va.: Time-Life Books, 1983.

Colón, Fernando, *The Life of the Admiral Christopher Columbus by His Son Ferdinand.* Transl. by Benjamin Keen. New Brunswick, N.J.: Rutgers University Press, 1959.

Cox, Allan, and Robert Brian Hart, *Plate Tectonics: How It Works.* Palo Alto, Calif.: Blackwell Scientific Publications, 1986.

Dott, Robert H., Jr., and Roger Lyman Batten, *Evolution of the Earth.* New York: McGraw-Hill, 1988.

Eicher, Don L., *Geologic Time.* Englewood Cliffs, N.J.: Prentice-Hall, 1976.

The Encyclopedia of Space Travel and Astronomy. New York: Crescent Books, 1985.

Fisher, David E., *The Birth of the Earth.* New York: Columbia University Press, 1987.

Gallant, Roy A., *Earth's Changing Climate.* New York: Four Winds Press, 1979.

Gillispie, Charles Coulston, *Dictionary of Scientific Biography.* New York: Charles Scribner's Sons, 1981.

Gleick, James, *Chaos: Making a New Science.* New York: Viking, 1987.

Glen, William, *The Road to Jaramillo.* Stanford, Calif.: Stanford University Press, 1982.

Gribbin, John, *Genesis: The Origins of Man and the Universe.* New York: Delta/Eleanor Friede, 1981.

Hallam, A., *Great Geological Controversies.* New York: Oxford University Press, 1983.

Henderson-Sellers, A., and P. J. Robinson, *Contemporary Climatology.* New York: Longman Scientific & Technical, 1986.

Horsfield, Brenda, and Peter Bennet Stone, *The Great Ocean Business.* New York: New American Library, 1972.

Kingston, Jeremy, and David Lambert, *Catastrophe and Crisis.* New York: Facts on File, 1979.

Leopold, Luna B., Kenneth S. Davis, and the Editors of Time-Life Books, *Water* (Life Science Library series).

Alexandria, Va.: Time-Life Books, 1980.

Lovelock, James, *The Ages of Gaia.* New York: W. W. Norton, 1988.

Mantell, Charles Letnam, *Our Fragile Water Planet.* New York: Plenum Press, 1976.

Margulis, Lynn, and Dorion Sagan, *Microcosmos.* New York: Summit Books, 1986.

Miller, Russell, and the Editors of Time-Life Books, *Continents in Collision* (Planet Earth series). Alexandria, Va.: Time-Life Books, 1983.

Oberg, James E., *New Earths.* Harrisburg, Pa.: Stackpole Books, 1981.

Philander, S. George, *El Niño, La Niña, and the Southern Oscillation.* London: Academic Press, 1989.

Schlee, Susan, *The Edge of an Unfamiliar World.* New York: E. P. Dutton, 1973.

Schmidt, Victor A., *Planet Earth and the New Geoscience.* Dubuque, Iowa: Kendall/Hunt, 1986.

Schneider, Stephen H., and Randi Londer, *The Coevolution of Climate and Life.* San Francisco: Sierra Club Books, 1984.

Scientific American, *Continents Adrift and Continents Aground.* San Francisco: W. H. Freeman, 1976.

Smith, David G., ed., *The Cambridge Encyclopedia of Earth Sciences.* Cambridge: Cambridge University Press, 1982.

Smith, Peter J., ed., *The Earth.* New York: Macmillan, 1986.

Snow, Theodore, P., *Essentials of the Dynamic Universe.* St. Paul: West, 1987.

Speth, James Gustave, "Environmental Pollution: A Long-Term Perspective." In *Earth '88: Changing Geographic Perspectives, Proceedings of the Centennial Symposium.* Washington, D.C.: National Geographic Society, 1988.

"Volcanoes and Climate." In *What's Happening in Chemistry?* Washington, D.C.: American Chemical Society, 1985.

Walker, James C. G., *Earth History: The Several Ages of the Earth.* Boston: Jones and Bartlett, 1986.

Weiner, Jonathan, *Planet Earth.* New York: Bantam Books, 1986.

Wilson, David, *Rutherford: Simple Genius.* London: Hodder and Stoughton, 1983.

Wood, Robert Muir, *The Dark Side of the Earth.* London: George Allen & Unwin, 1985.

Young, Louise B., *Earth's Aura.* New York: Alfred A. Knopf, 1977.

Periodicals

Abelson, Philip H.:
"Air Pollution and Acid Rain." *Science,* November 8, 1985.
"Climate and Water." *Science,* January 27, 1989.
"Greenhouse Role of Trace Gases." *Science,* March 14, 1986.

Alexander, Tom, "A Revolution Called Plate Tectonics Has Given Us a Whole New Earth." *Smithsonian,* January 1975.

Allman, William F., "Rediscovering Planet Earth." *U.S. News & World Report,* October 31, 1988.

Anderson, Don L., and Adam M. Dziewonski, "Seismic Tomography." *Scientific American,* October 1984.

Asimov, Isaac, "Water, Water Everywhere But . . . ." *International Wildlife,* March/April 1978.

Bambach, Richard K., Christopher R. Scotese, and Alfred M. Ziegler, "Before Pangea: The Geographies of the Paleozoic World." *American Scientist,* January/February 1980.

Bartusiak, Marcia, "Mapping the Sea Floor from Space." *Popular Science,* February 1984.

Batisse, Michel, "The Strange Ways of H_2O." *Unesco Courier,* February 1978.

Bergman, Kenneth H., Alan D. Hecht, and Stephen H. Schneider, "Climate Models." *Physics Today,* October 1981.

Bernardo, Stephanie, "The Seafloor: A Clear View from Space." *Science Digest,* June 1984.

Berner, Robert A., and Antonio C. Lasaga, "Modeling the Geochemical Carbon Cycle." *Scientific American,* March 1989.

Broecker, Wallace S.:
"The Biggest Chill." *Natural History,* October 1987.
"The Ocean." *Scientific American,* September 1983.

Brush, Stephen G.:
"Cooling Spheres and Accumulating Lead: The History of Attempts to Date the Earth's Formation." *The Science Teacher,* December 1987.
"Inside the Earth." *Natural History,* February 1984.

Bryan, K., S. Manabe, and M. J. Spelman, "Interhemispheric Asymmetry in the Transient Response of a Coupled Ocean-Atmosphere Model to a CO_2 Forcing." *Journal of Physical Oceanography,* June 1988.

Bryan, K., et al., "Transient Climate Response to Increasing Atmospheric Carbon Dioxide." *Science,* January 1, 1982.

Burchfiel, B. Clark, "The Continental Crust." *Scientific American,* September 1983.

Canby, Thomas Y., "El Niño's Ill Wind." *National Geographic,* February 1984.

Carey, John, "Are Deserts on the March?" *International Wildlife,* March/April 1985.

Carroll, Louise Purrett, "The Paleomap Project." *The World & I,* March 1989.

Carter, William E., and Douglas S. Robertson, "Studying the Earth by Very-Long-Baseline Interferometry." *Scientific American,* November 1986.

Cordell, Bruce M., "Mars, Earth, and Ice." *Sky & Telescope,* July 1986.

Covey, Curt, "The Earth's Orbit and the Ice Ages." *Scientific American,* February 1984.

Cox, Allan, G. Brent Dalrymple, and Richard R. Doell, "Reversals of the Earth's Magnetic Field." *Scientific American,* February 1967.

Cromie, William J., "The New Oceanography." *Mosaic,* July/August 1980.

Davis, L., "Biological Diversity: Going . . . Going . . . ?" *Science News,* September 27, 1986.

"Dinosaurs and the Great Extinction." *Science Impact Letter,* May 1989.

Eberhart, Jonathan:
"And into the Warming Sea Rode the 4,000." *Science News,* June 1, 1974.
"The GATE Researchers in the Tropics Obtained Their Data—and a New Image for 'Big Science.' " *Science News,* November 23, 1974.

Ellis, William S.:
"Africa's Sahel: The Stricken Land." *National Geographic,* August 1987.
"Brazil's Imperiled Rain Forest." *National Geographic,* December 1988.

Francheteau, Jean, "The Oceanic Crust." *Scientific American,* September 1983.

Gloersen, Per, and William J. Campbell, "Variations in the Arctic, Antarctic, and Global Sea Ice Covers during 1978-1987 as Observed with the Nimbus 7 Scanning Multichannel Microwave Radiometer." *Journal of Geophysical Research,* September 15, 1988.

"The Greenhouse Effect: How It Can Change Our Lives." *EPA Journal,* January/February 1989.

Hammond, Allen L., "Plate Tectonics: The Geophysics of the Earth's Surface." *Science,* July 2, 1971.

Hansen, J., et al., "Global Climate Changes as Forecast by Goddard Institute for Space Studies Three-Dimensional Model." *Journal of Geophysical Research,* August 20, 1988.

Hays, J. D., John Imbrie, and N. J. Shackleton, "Variations in the Earth's Orbit: Pacemaker of the Ice Ages." *Science,* December 10, 1976.

"Heat Beat." *Scientific American,* August 1985.

Heirtzler, J. R., "Sea-Floor Spreading." *Scientific American,* December 1968.

Hekinian, Roger, "Undersea Volcanoes." *Scientific American,* July 1984.

Heppenheimer, T. A., "Journey to the Center of the Earth." *Discover,* November 1987.

Hilts, Philip J.:
"Answers from Deep Inside the Planet." *Washington Post,* June 5, 1988.
"Some 'Greenhouse' Effects: Pestilence, Super-Storms?" *Washington Post,* December 8, 1988.

Hofmann, D. J., "Direct Ozone Depletion in Springtime Antarctic Lower Stratospheric Clouds." *Nature,* February 2, 1989.

Houghton, Richard A., and George M. Woodwell, "Global Climatic Change." *Scientific American,* April 1989.

Hurley, Patrick M., "The Confirmation of Continental Drift." *Scientific American,* April 1968.

Imbrie, J., and J. Z. Imbrie, "Modeling the Climatic Response to Orbital Variations." *Science,* February 29, 1980.

Ingersoll, Andrew P., "The Atmosphere." *Scientific American,* September 1983.

Jeanloz, Raymond, "The Earth's Core." *Scientific American,* September 1983.

Kasting, James F., Owen B. Toon, and James B. Pollack, "How Climate Evolved on the Terrestrial Planets." *Scientific American,* February 1988.

Kellogg, W. W., and S. H. Schneider, "Climate Stabilization: For Better or for Worse?" *Science,* December 27, 1974.

Kerr, Richard A.:
"Capturing El Niño in Models." *Science,* December 11, 1987.
"Carbon Dioxide and Climate: Carbon Budget Still Unbalanced." *Science,* September 30, 1977.
"Climate Since the Ice Began to Melt." *Science,* October 19, 1984.
"Research News: The Global Warming Is Real." *Science,* February 3, 1989.
"How Much Drying from a Greenhouse Warming?" *Science,* January 23, 1987.
"How to Fix the Clouds in Greenhouse Models." *Science,* January 6, 1989.
"Making the World's Roof." *Science,* May 22, 1987.
"The Mantle's Structure—Having It Both Ways." *Science,* June 24, 1988.

Kiester, Edwin, Jr., "A Deathly Spell Is Hovering above the Black Forest." *Smithsonian,* November 1985.

"Larger El Niños?" *Science Impact,* November 1987.

Lemonick, Michael D., "Journey to the Earth's Core." *Time,* June 13, 1988.

Levy, Hiram, II, and Walter J. Moxim, "Influence of Long-Range Transport of Combustion Emissions on the Chemical Variability of the Background Atmosphere." *Nature,* March 23, 1989.

"Living in the Greenhouse." *The Economist,* March 11, 1989.

Luoma, Jon R., "Forests Are Dying But Is Acid Rain Really to Blame?" *Audubon,* March 1987.

McKenzie, D. P., "The Earth's Mantle." *Scientific American,* September 1983.

MacLeish, William H.:
"The Blue God." *Smithsonian,* February 1989.
"Painting a Portrait of the Stream from Miles Above—and Below." *Smithsonian,* March 1989.

Malone, Thomas F., "Mission to Planet Earth." *Environment,* October 1986.

Manabe, S., and Kirk Bryan, "Climate Calculations with a Combined Ocean-Atmosphere Model." *Journal of the Atmospheric Sciences,* July 1969.

Manabe, S., and R. J. Stouffer, "Two Stable Equilibria of a Coupled Ocean-Atmosphere Model." *Journal of Climate,* September 1988.

Manabe, S., and R. T. Wetherald, "Reduction in Summer Soil Wetness Induced by an Increase in Atmospheric Carbon Dioxide." *Science,* May 2, 1986.

Maranto, Gina, "Are We Close to the Road's End?" *Discover,* January 1986.

Matson, Michael, "The 1982 El Chichón Volcano Eruptions—a Satellite Perspective." *Journal of Volcanology and Geothermal Research,* Vol. 23, 1984, pages 1-10.

Matthews, Samuel W., "What's Happening to Our Climate?" *National Geographic,* November 1976.

"Milankovitch Climate Cycles: Old and Unsteady." *Science,* September 4, 1981.

Molina, Mario J., et al., "Antarctic Stratospheric Chemistry of Chlorine Nitrate, Hydrogen Chloride, and Ice: Release of Active Chlorine." *Science,* November 27, 1987.

Molnar, Peter, "The Structure of Mountain Ranges." *Scientific American,* July 1986.

Molnar, Peter, and Paul Tapponnier, "The Collision between India and Eurasia." *Scientific American,* April 1977.

Monastersky, Richard:
"Clouds without a Silver Lining." *Science News,* October 15, 1988.
"Decline of the CFC Empire." *Science News,* April 9, 1988.
"Delving Deep into the Indian Past." *Science News,* July 25, 1987.

"The Plankton-Climate Connection." *Science News,* December 5, 1987.

"Quake Prediction: Magnetic Signals?" *Science News,* September 12, 1987.

"Shrinking Ice May Mean Warmer Earth." *Science News,* October 8, 1988.

"The Whole-Earth Syndrome." *Science News,* June 11, 1988.

Morelli, Andrea, and Adam M. Dziewonski, "Topography of the Core-Mantle Boundary and Lateral Homogeneity of the Liquid Core." *Nature,* February 19, 1987.

Murphy, Jamie, "The Quiet Apocalypse." *Time,* October 13, 1986.

Nance, R. Damian, Thomas R. Worsley, and Judith B. Moody, "The Supercontinent Cycle." *Scientific American,* July 1988.

Olson, Steve, "Computing Climate." *Science 82,* May 1982.

Peters, Robert L., and Joan D. S. Darling, "The Greenhouse Effect and Nature Reserves." *BioScience,* December 1985.

Plass, Gilbert N., "Carbon Dioxide and Climate." *Scientific American,* July 1959.

Postel, Sandra, and Lori Heise, "The Fragile Forest." *The Courier,* January 1989.

"The Race to Predict Next Week's Weather." *Science,* April 1, 1983.

Raloff, J., "Ozone Hole of 1988: Weak and Eccentric." *Science News,* October 22, 1988.

Ramanathan, V., "The Greenhouse Theory of Climate Change: A Test by an Inadvertent Global Experiment." *Science,* April 15, 1988.

Ramanathan, V., et al., "Cloud-Radiative Forcing and Climate: Results from the Earth Radiation Budget Experiment." *Science,* January 6, 1989.

Rasmussen, R. A., and M. A. K. Khalil, "Atmospheric Trace Gases: Trends and Distributions over the Last Decade." *Science,* June 27, 1986.

Revelle, Roger, "Carbon Dioxide and World Climate." *Scientific American,* August 1982.

Roberts, Leslie, "Is There Life after Climate Change?" *Science,* November 18, 1988.

Schlee, Susan, "Science and the Sea." *The Wilson Quarterly,* summer 1984.

Schneider, Stephen H.:
"Climate Modeling." *Scientific American,* May 1987.
"The Greenhouse Effect: Science and Policy." *Science,* February 10, 1989.

Shabecoff, Philip, "Global Warming: Experts Ponder Bewildering Feedback Effects." *New York Times,* January 17, 1989.

Shaw, Robert W., "Air Pollution by Particles." *Scientific American,* August 1987.

"Ships' Smokestacks Emerge as Factor in Earth's Climate." *New York Times,* September 1, 1987.

Simmons, Marlise, "Ambitious Amazon Expedition Probes Forest's Role in Global Atmosphere." *New York Times,* June 2, 1987.

Simon, C., "Modeling India's Drive into Eurasia." *Science News,* December 18 and 25, 1987.

Stolarski, Richard S., "The Antarctic Ozone Hole." *Scientific American,* January 1988.

Strickland, Edwin, "The Colors of Mars." *Astronomy,* May 1980.

Tapponnier, Paul:
"La Grand Choc de l'Inde et de la Chine." *Geo,* February 1986.
"A Tale of Two Continents." *Natural History,* November 1986.

Thayer, Victoria G., and Richard T. Barber, "At Sea with El Niño." *Natural History,* October 1984.

Thompson, Lonnie G., Stefan Hastenrath, and Benjamin Morales Arnao, "Climatic Ice Core Records from the Tropical Quelccaya Ice Cap." *Science,* March 23, 1979.

Thompson, Lonnie G., et al., "A 1500-Year Record of Tropical Precipitation in Ice Cores from the Quelccaya Ice Cap, Peru." *Science,* September 6, 1985.

Tilling, Robert, "Volcanic Cloud May Alter Earth's Climate." *National Geographic,* November 1982.

Toon, Owen B., and Steve Olson, "The Warm Earth." *Science 85,* October 1985.

Vink, Gregory E., W. Jason Morgan, and Peter R. Vogt, "The Earth's Hot Spots." *Scientific American,* April 1985.

Voropayev, Grigori, "Water and Man." *The Courier,* January 1985.

Wanning, Esther, "Interview: Roger Revelle." *Omni,* March 1984.

Weintraub, Boris, "Fire and Ash, Darkness at Noon." *National Geographic,* November 1982.

Weisburd, Stefi:
"How Hot Is the Heart of the Earth?" *Science News,* April 18, 1987.
"Plunging Plates Cause a Stir." *Science News,* August 16, 1986.
"Rooting for Continental Roots." *Science News,* December 13, 1986.
"Seismic Journey to the Center of the Earth." *Science News,* July 5, 1986.

Wesson, Robert L., and Robert E. Wallace, "Predicting the Next Great Earthquake in California." *Scientific American,* February 1985.

Wiebe, Peter, "Rings of the Gulf Stream." *Scientific American,* March 1982.

Wilson, J. Tuzo, "Evidence from Islands on the Spreading of Ocean Floors." *Nature,* February 9, 1963.

Yulsman, Tom, "Plate Tectonics Revised." *Science Digest,* November 1985.

Other Sources

"Airborne Arctic Stratospheric Expedition: Preliminary Findings." Press statement. National Oceanic and Atmospheric Administration, February 17, 1989.

"The Blue Planet: Physical and Chemical Makeup of the Oceans and Dynamics of the Oceans." Video from the "Planet Earth" television series. Pittsburgh: WQED/ Pittsburgh, 1986.

Brush, Stephen G., and S. K. Banerjee, "Geomagnetic Secular Variation and Theories of the Earth's Interior." Paper presented at the meeting of the International Union of Geodesy and Geophysics, Vancouver, British Columbia, August 15, 1987.

Burch, Susan, "Acid Rain May Boost Mosses as Tree Killers Worldwide." News release. Boulder, Colo.: National Center for Atmospheric Research, September 9, 1988.

Camilliere, June, "Methane Source Sheds Light on Greenhouse Effect." News release. Boulder, Colo.: National Center for Atmospheric Research, February 14, 1989.

"The Climate Puzzle: The Atmosphere and Climates of Earth." Video from the "Planet Earth" television series. Pittsburgh: WQED/Pittsburgh, 1986.

"The Crucial Decade: The 1990s and the Global Environmental Challenge." Policy paper. World Resources Institute, January 1989.

"Earth System Science: A Closer View." NASA Advisory Council report. Washington, D.C.: NASA, Earth System Sciences Committee, January 1988.

"Fate of the Earth: The Balance of Nature and the Impact of Man." Video from the "Planet Earth" television series. Pittsburgh: WQED/Pittsburgh, 1986.

Hansen, J., et al., "Regional Greenhouse Climate Effects." Reprinted from *Preparing for Climate Change: Proceedings of the Second North American Conference on Preparing for Climate Change, December 6-8, 1988.* Washington, D.C.: Climate Institute, 1989.

"The Living Machine: Plate Tectonics and Continental Tectonics and the Earth's Interior." Video from the "Planet Earth" television series. Pittsburgh: WQED/Pittsburgh, 1986.

"A Preview of Planet Earth." Pittsburgh: Metropolitan Pittsburgh Public Broadcasting, 1985.

Spelman, M., S. Manabe, and K. Bryan. "Transient Climate Change Due to Increasing Greenhouse Gases." Unpublished paper, February 1989.

INDEX

Mexico: El Chichón ash cloud, *82*
Midocean ridges, 45-46, 47, 58; in carbon cycle, *105;* and fracture zones, 50-51; Pacific-Antarctic, data from, 52-53
Milankovitch, Milutin, 74-75
Milne, John, 36
Mitchell, Mount, N.C.: tree death, 122
Modeling of climate, 82-84, *126-129*
Mohorovičić, Andrija, 37, 40
Mohorovičić discontinuity ("Moho"), 40, 45
Monsoon rains, failure of, 80, 94
Moon: earthrise over, *27, 29, 31, 33*

N
Namias, Jerome: quoted, 81
National Aeronautics and Space Administration (NASA) climate research, 98-99; Airborne Antarctic Ozone Experiment, 124-125; Goddard Institute for Space Studies models, *126, 127,* 128
Nazca Plate subduction, *60, 61*
Neptunist theory of Earth, 26
Nile (river), Africa, *116-117*
Nitrogen in Earth's atmosphere, 14
Nitrous oxides, 114-115
Numerical models, 82-84, *126-129*

O
Ocean, 22; basin formation, *58;* in carbon cycle, *104-105;* clouds over, *66-67, 72, 132-133;* convection currents from, 90, *91;* currents, 69, *70-71,* 79-80, *92-93;* early, 16, 18, 26; El Niño and La Niña conditions, 79-81, 83, *94-95,* 99; evaporation from, *88-89;* plankton, *74,* 75, *106,* 107; pollution over, *120-121;* transition zone, 78, 86; winds over, *66-67, 91. See also* Marine geophysics
Oldham, Richard Dixon, 37, 40
Orbital variations, Earth's, and ice ages, 74-75, *76-77*
Owen, H.M.S., 48
Oxygen: in foraminifera shells, 75; in history of Earth, 18, 130; ozone, 18, 86, 115, 122-126, *124-125;* in water molecule, *24*
Ozone as greenhouse gas, 115
Ozone layer, 18, 86, 122-123; depletion of, 123-126, *124-125*

P
Pacific-Antarctic ocean ridge, 52-53
Pacific Ocean: clouds over, *66-67, 72;* El Niño and La Niña conditions, 79-81, 83, *94-95,* 99; gyres, *92-93*
Pacific Plate, 60, 62, 64
Paleomagnetism, 43; pattern of

variation, 48-49, *51,* 52-53; reversals of polarity, 43, 49, 51-52
Pangaea (hypothetical supercontinent), 41-42, *46-47*
Patterson, Clair, 33
Peru: El Niño and La Niña conditions, 79-81, *94-95,* 99; icecap, *81*
Peru-Chile Trench: formation of, *60*
Photons: chlorine atoms freed by, *124, 125*
Photosynthesis, 18, 130
Pitman, Walter, III, 52-53
Planets, 22; formation of, *6-7*
Plankton, *74,* 75, *106,* 107
Plate tectonics. *See* Tectonic plates
Pollack, James, 107
Precession of equinoxes, *77*
Precipitation changes: modeled, *127,* 129
P-waves, 37, *38, 39,* 40, 41

Q
Quelccaya icecap, Peru, *81*

R
Radiation: solar, *86-87,* 101, *102-103*
Radioactive decay: as dating tool, 32-33; discovery of, 30; and half-life calculations, 32; heat from, 30-31; in history of Earth, 11, 31-33
Radiolarians, *74,* 75
Radium, 30
Rain: acid, 119, 122; El Niño vs. monsoon, 80, 81, 94, *95;* formation of, *89;* in history of Earth, 14, 16; reduced, deforestation and, 114
Rain forest: clearing, effects of, 112, *113,* 114
Rennell, John: quoted, 70
Richardson, Lewis F., 73
Ridges: in carbon cycle, *105;* and fracture zones, 50-51; midocean, 45-46, 47, 58; Pacific-Antarctic, data from, 52-53
Rifts and rifting, *45,* 46, *58-59*
Rotation of Earth: axis of, and ice ages, 74-75, *76-77;* and Coriolis effect, *90;* and ocean currents, 92
Rutherford, Ernest, 30, 31, 32; quoted, 31-32

S
Sahara (desert), Africa: advance of, 115, *116-117,* 118
Sahel (region), Africa, 115, 118
San Andreas fault, Calif., 51, 62
São Paulo, Brazil: smoke and haze over, *110-111*
Satellite images: African desertification, *116-117;* Antarctic sea ice, *78-79;* El Chichón ash cloud, *82;* Gulf Stream, *70-71*

Schneider, Stephen, 108; quoted, 84, 108
Seafloor spreading, 46, 47-48, 50-51, 53, *58;* and magnetization pattern, 49, *51,* 53
Seamounts (guyots), 47
Seasonal variation: in Antarctic sea ice, *78-79;* in carbon dioxide, 109
Sediments, marine, 44-45; climate history recorded in, *74,* 75, 78
Seismic prospecting, 44
Seismic tomography, 54
Seismic waves and velocities, 36-37, *38, 39,* 40, 41, 54
Seismographs, 36, 37, 38; nuclear tests detected by, 50
Sensitive dependence, 84
Shackleton, Nicholas, 75, 78
Shadow zones, seismic, 37, *38,* 40, 41
Shoemaker, Eugene, 131
Smog, 108, 119
Smoke over South America, *110-111, 113*
Soddy, Frederick, 30, 32
Soil moisture: modeled changes in, *129*
Solar System: formation of, *6-7*
Sonar, 43-44
South America: Amazon area, 98, 112, *113,* 114; El Niño and La Niña conditions, 79-81, *94-95,* 99; smoke over, *110-111, 113*
South American Plate: subduction by, *60, 61*
Southern oscillation, 80, 94
Space shuttle, views from: Amazon River basin, *113;* Himalayas, *34-35*
Stratocumulus clouds, *72*
Stratosphere, *86-87;* ozone layer, 18, 86, 122-126, *124-125*
Strutt, Robert J., 30-31
Subduction, 53, *60-61, 105*
Sulfate aerosols, 119
Sun: in history of Earth, *6-7,* 18, 108; Kelvin's theory of, 28-29; radiation from, *86-87,* 101, *102-103;* unequal heating of Earth by, 90
S-waves, 37, *38, 39,* 40
Sykes, Lynn, 50

T
Tectonic plates, 16, 53, 55, *57;* and climate change, 100; Himalayas as result, *34-35, 52;* rifting, *45,* 46, *58-59;* subduction, 53, *60-61, 105;* time line, *44-49;* transform faulting, 51, *62-63;* volcanic islands in middle, 50, *64-65;* in Wegener's theory, 41-42
Tharp, Marie, 45-46
Thermocline, 78
Thermodynamics, laws of: and age of Earth, 29-30

ACKNOWLEDGMENTS

The editors wish to thank Philip Ardanuy, Research and Data Systems Co., Greenbelt, Md.; Stephen Brush, University of Maryland, College Park, Md.; Ben Chao, NASA Goddard Space Flight Center, Greenbelt, Md.; James Hansen, NASA Goddard Institute for Space Studies, New York, N.Y.; Joachim Kuettner, NCAR, Boulder, Colo.; Gerry Livingston, Ames Research Center, Moffett Field, Calif.; Syukuro Man-abe, Princeton University, Princeton, N.J.; Denise Manning, Annenberg/CPB, Washington, D.C.; Leslie Morrissey, Ames Research Center, Moffett Field, Calif.; S. George Philander, NOAA, Princeton, N.J.; David Rubincam, NASA Goddard Space Flight Center, Greenbelt, Md.; Michael Turner, Freshwater Institute, Winnipeg, Manitoba, Canada; Lisa Vasquez-Morrison, Lyndon B. Johnson Space Center, Houston, Tex.

PICTURE CREDITS

The sources for the illustrations that appear in this book are listed below. Credits from left to right are separated by semicolons, from top to bottom by dashes.

Cover: NASA, LBJ Space Center (AS17-148-22727). Front and back endpapers: Artwork by Time-Life Books. 6-19: Art by Matt McMullen. 20, 21: NASA, Washington, D.C. (72-HC-463). 22: Initial cap, detail from pages 20, 21. 24: Art by Fred Holz. 27: NASA, LBJ Space Center (AS11-44-6547). 29: NASA, LBJ Space Center (AS11-44-6549). 31: NASA, LBJ Space Center (AS11-44-6550). 33: NASA, LBJ Space Center (AS11-44-6552). 34, 35: NASA, LBJ Space Center (41G-12G-009). 36: Initial cap, detail from pages 34, 35. 38, 39: Art by Yvonne Gensurowsky/Stansbury, Ronsaville, Wood, Inc. 43-52: Art by Fred Holz. 56-65: Art by Alfred Kamajian. 66, 67: NASA, Washington, D.C. (69-HC-199). 68: Initial cap, detail from pages 66, 67. 70, 71: Courtesy O. Brown and R. Evans, University of Miami. 72: NASA, LBJ Space Center (AS6-2-1425). 74: Stanley A. Kling, Micropaleo Consultants, Inc., Encinitas, Calif. 76, 77: Art by Mark Robinson. 78, 79: Per Gloersen, NASA, Goddard Space Flight Center. 81: Lonnie G. Thompson, Byrd Polar Research Center, Ohio State University. 82: NOAA/NESDIS. 85-94: Art by Rob Wood/Stansbury, Ronsaville, Wood, Inc. 95: Art by Rob Wood/SRW, Inc.; photos, R. Legeckis/NOAA (3). 96, 97: NASA, LBJ Space Center (AS12-51-7587). 98: Initial cap, detail from pages 96, 97. 102, 103: Art by Yvonne Gensurowsky/SRW, Inc. 104, 105: Art by Alfred Kamajian. 106: NASA, LBJ Space Center (61A-50-020). 110, 111: © Nicholas de Vore III/Bruce Coleman, Inc. 113: NASA, LBJ Space Center (STS26-38-014T); inset, NASA, LBJ Space Center (SL2-05-325). 114, 115: Art Wolfe, Inc. 116, 117: P. Jacobberger, Center for Earth and Planetary Studies, National Air and Space Museum (2); Earth Satellite Corporation, Chevy Chase, Md. 120, 121: NASA, LBJ Space Center (51G-47-084). 123: Fredrik Ehrenstrom, Oxford Scientific Films, Long Hanborough, Oxfordshire. 124, 125: Art by Yvonne Gensurowsky/SRW, Inc. 126-129: Artwork by Time-Life Books. 132, 133: NASA, LBJ Space Center (51G-44-091).

Time-Life Books Inc.
is a wholly owned subsidiary of
TIME INCORPORATED

Editor-in-Chief: Jason McManus
Chairman and Chief Executive Officer:
J. Richard Munro
President and Chief Operating Officer:
N. J. Nicholas, Jr.
Editorial Director: Richard B. Stolley

THE TIME INC. BOOK COMPANY
President and Chief Executive Officer:
Kelso F. Sutton
President, Time Inc. Books Direct:
Christopher T. Linen

TIME-LIFE BOOKS INC.
EDITOR: George Constable
Executive Editor: Ellen Phillips
Director of Design: Louis Klein
Director of Editorial Resources: Phyllis K. Wise
Editorial Board: Russell B. Adams, Jr., Dale M.
Brown, Roberta Conlan, Thomas H. Flaherty, Lee
Hassig, Donia Ann Steele, Rosalind Stubenberg
Director of Photography and Research:
John Conrad Weiser
Assistant Director of Editorial Resources:
Elise Ritter Gibson

PRESIDENT: John M. Fahey, Jr.
Senior Vice Presidents: Robert M. DeSena, James
L. Mercer, Paul R. Stewart, Joseph J. Ward
Vice Presidents: Stephen L. Bair, Stephen L.
Goldstein, Juanita T. James, Andrew P. Kaplan,
Carol Kaplan, Susan J. Maruyama, Robert H.
Smith
Supervisor of Quality Control: James King

PUBLISHER: Joseph J. Ward

Editorial Operations
Copy Chief: Diane Ullius
Production: Celia Beattie
Library: Louise D. Forstall

Correspondents: Elisabeth Kraemer-Singh (Bonn);
Christina Lieberman (New York); Maria Vincenza
Aloisi (Paris); Ann Natanson (Rome). Valuable
assistance was also provided by Christine Hinze
(London).

VOYAGE THROUGH THE UNIVERSE

SERIES DIRECTOR: Roberta Conlan
Series Administrator: Judith W. Shanks

Editorial Staff for *The Third Planet*
Designer: Dale Pollekoff
Associate Editor: Sally Collins (pictures)
Text Editors: Carl Posey (principal), Pat Daniels
Researchers: Patti H. Cass, Barbara C. Mallen,
Edward O. Marshall
Writer: Esther Ferington
Assistant Designer: Brook Mowrey
Copy Coordinator: Darcie Conner Johnston
Picture Coordinator: Ruth Moss
Editorial Assistant: Jayne A. L. Dover

Special Contributors: Peter Gwynne, Allen
Hammond, Richard Kerr, Chuck Smith, Mark
Washburn, Robert White (text); Sydney Baily,
Susan Bender, Adam Dennis, Jocelyn Lindsay,
Jacqueline Shaffer, Joann Stern (research);
Barbara L. Klein (index).

CONSULTANTS
KENNETH H. BERGMAN is the senior program officer of the Board on Atmospheric Sciences and Climate of the National Academy of Sciences. He specializes in meteorology and climatology.

JAMES R. HEIRTZLER, director of geophysics,
NASA Goddard Space Flight Center, is a specialist in
earth sciences, particularly regarding the theory of
plate tectonics and seismology.

RONALD B. PARKER taught geology at the University of Wyoming for seventeen years. He has received numerous grants and written over fifty ar-
ticles on earth science.

S. ICHTIAQUE RASOOL is the chief scientist for
Global Change Programs in the Office of Space Science and Applications at NASA Headquarters. His
main research interest is the field of physics of
atmospheres.

BRUCE R. ROSENDAHL is currently a professor of
geophysics at Duke University. He specializes in the
study of rifted plate margins using exploration seismic methods.

STEPHEN H. SCHNEIDER, head of the Interdisci-
plinary Climate Systems section at the National
Center for Atmospheric Research, is interested in
climatic change, climatic modeling of paleocli-
mates, and human impacts on climate.

CHRISTOPHER R. SCOTESE is a research geologist
for Shell Development Company and the chairman
of the Paleomap Project of the International Lithosphere Project.

**Library of Congress Cataloging in
Publication Data**
The Third planet/by the editors of
Time-Life Books.
p.   cm.   (Voyage through the universe).
Bibliography: p.
Includes index.
ISBN 0-8094-6879-4
ISBN 0-8094-6880-8 (lib. bdg.)
1. Earth—Popular works. I. Time-Life Books.
II. Series.
QB631.2.T48 1989
525—dc20   89-4467 CIP

For information on and a full description of
any of the Time-Life Books series, please call
1-800-621-7026 or write:
Reader Information
Time-Life Customer Service
P.O. Box C-32068
Richmond, Virginia 23261-2068

Earth: diameter 7,926 miles

Neptune: diameter 30,200 miles

Uranus: diameter 31,600 miles

Red supergiant: diameter 400 million miles

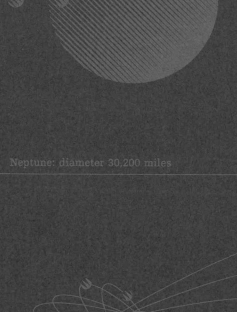

Solar System: diameter 7.5 billion miles

Globular cluster: diameter 2×10^{14} miles

Milky Way: diameter 100,000 light-years

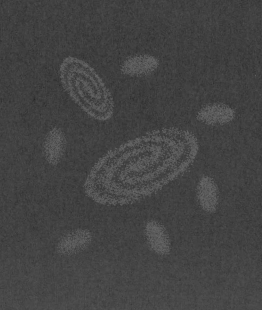

Local Group of galaxies:
6 million light-years across

Largest double radio source:
length 17 million light-years